NEEDLELESS ELECTROSPINNING OF NANOFIBERS

Tong Lin
Xungai Wang

NEEDLELESS ELECTROSPINNING OF NANOFIBERS

Technology and Applications

PAN STANFORD PUBLISHING

Published by

Pan Stanford Publishing Pte. Ltd.
Penthouse Level, Suntec Tower 3
8 Temasek Boulevard
Singapore 038988

Email: editorial@panstanford.com
Web: www.panstanford.com

British Library Cataloguing-in-Publication Data
A catalogue record for this book is available from the British Library.

Needleless Electrospinning of Nanofibers:
Technology and Applications

ISBN 978-981-4316-35-4 (Hardcover)
ISBN 978-981-4316-84-2 (eBook)

Printed in the USA

Contents

Preface

Electrospun nanofibers have shown enormous application potential in diverse areas. However, most works on electrospinning nanofibers are based on conventional needle electrospinning techniques, and mass production of nanofibers for practical applications has been a great challenge with the needle electrospinning systems. This book systemically introduces a relatively new electrospinning process, needleless electrospinning. The book is divided into seven chapters. In the first chapter, basic electrospinning theory is introduced. The second chapter summarizes the functional applications of electrospun nanofibers in different fields. Chapter 3 reviews the emerging electrospinning technologies that have potential for large-scale production of nanofibers. Chapters 4 and 5 describe an important needleless electrospinning technique using different fiber generators such as ball, cylinder, disc, and wire coil, and the effects of the fiber generator, its shape and dimension, and operating parameters on electrospinning performance, fiber morphology, and productivity are also introduced. Chapter 6 analyzes the electric field profiles and provides a method to calculate the electric field intensity in an electrospinning zone. Chapter 7 discusses the influence of the fiber collector on fiber quality. The book will be a very useful resource for academics, industry professionals, and graduate students in nanomaterials, nanostructures, and nanofabrication .

Acknowledgments

This book is mainly based on the research results of three PhD students, Dr. Jian Fang, Dr. Haitao Niu, and Dr. Xin Wang, from the Institute for Frontier Materials (IFM) at Deakin University, Australia. The three students completed their PhD projects under the supervision of the authors of this book. This book would not have been possible without their dedicated studies and support. We are also indebted to Mr. Graeme Keating and Dr. Chris Hurren, who assisted in the vital tasks of constructing the prototype needleless electrospinning setups in our lab. Their invaluable contributions to this book are acknowledged by the authors. The authors also thank the funding support from the Australian Research Council under its Discovery Project scheme, and Deakin University under its Central Research Grant scheme. Without the funding support, the experiments would not have been performed.

Chapter 1

Introduction to Electrospinning

Nanofibers are defined as one-dimensional nanomaterials with diameters less than 1 μm (1000 nm), and an aspect ratio (length/diameter) larger than 100:1. They are also named as superfine or ultrathin fibers in some literature [1, 2]. When the fibers are in the range of 100–1000 nm, they are also referred to as submicron fibers [3–5]. This intrinsic feature offers them drastically increased surface-to-volume ratio and high aspect ratio. There are several methods to produce nanofibers, including phase separation [6], self-assembly [7, 8], template synthesis [9], melt-blowing [10], flash spinning [11], bicomponent spinning [12], and electrospinning [13, 14]. Nanofiber mats produced from electrospinning have a naturally formed porous structure with excellent pore interconnectivity, and the pores are in the range between tens of nanometers to a few micrometers. The open pore structure and high permeability to gas, along with the high surface area, make them ideal porous membranes. Compared with other one-dimensional nanostructures (e.g., nanotubes or nano-rods), continuous nanofibers are advantageous in terms of fabrication cost and the possibility of being integrated into other desired assembly in one step.

1.1 Electrospinning

Electrospinning has become a very important nano-fabrication technology because of its versatility in processing a wide range

Needleless Electrospinning of Nanofibers: Technology and Applications
Tong Lin and Xungai Wang

ISBN 978-981-4316-35-4 (Hardcover), 978-981-4316-84-2 (eBook)
www.panstanford.com

of polymers, and the consistency in producing nanofibers with controllable fiber diameter and morphology. Electrospinning has been proven to be an ideal method of producing continuous nanofibers. This technique involves stretching a polymer solution (or melt) under a high electric field to form solid filaments, which deposit on the counter electrode forming a randomly orientated nanofiber web in most cases. Electrospinning has shown the following advantages:

- The setup is easy to build
- Fiber diameter is controllable
- Versatile fiber morphologies
- Nanofibers can be prepared from a wide range of materials
- Tailored structures (both 2D and 3D)
- Large-scale production potential

1.1.1 Brief History

Electrospinning stemmed from research on charged liquid in an electric field. William Gilbert first described the observation of a liquid interacting with electrostatic force in 1600 [15]. He found that a spherical water drop on a dry surface was drawn up into a cone when a piece of rubbed amber was held at a suitable distance above. Gilbert explained that charges were induced to form on the drop surface by the electric field, which resulted in electrostatic forces opposing the surface tension of the liquid. Based on this finding, electrospraying was first carried out to produce aerosol in 1745.

Nearly 300 years after the Gilbert's observation, Lord Rayleigh [16, 17] calculated the minimum charge for a drop to overcome its surface tension and create a liquid jet under an electric field. In the early 20th century, Zeleny [18] studied the behavior of fluid droplets at the end of metal capillaries and evaluated surface charge intensity by reading pressure change in a liquid reservoir. Some very important patents, which are regarded as the real beginning of electrospinning, were then filed by Formhals [19–21]. In his patents, Formhals proposed a new device to produce continuous fine fibers. About 20 years later, Taylor in 1960s studied the shape of cone formed by a fluid droplet under the action of an electric field and concluded that the conical interface between air and fluid was stable at the semi-angle conical angle of 49.3° [22]. The formation of such a cone structure in electrospraying and electrospinning has since been named as "Taylor cone." Taylor [23] also suggested that in a parallel electrical field, an electrically driven water jet had two critical instabilities: Rayleigh and bending instability. The Rayleigh instability is due to the surface tension, which is always trying to minimize surface area and break the jet into droplets, while the electrical bending instability is trying to bend the jet into spiral coils. For an electrically driven viscous fluid, the first instability can be suppressed by the viscous stress of the fluid, and the second instability is from the competition between the viscous stress along the jet and the electrostatic force working on the jet. When the electrostatic force is higher, it overcomes the viscous stress and elongates the jet.

Since Taylor's research, only a few limited improvements had been made in electrospinning, which did not attract much interest until Doshi and Reneker [13, 24] reported their work on using this technique to fabricate nano-structured materials in 1993. Since then, numerous studies have been conducted on electrospinning, and these studies have covered the theories and processes of electrospinning and the applications of electrospun nanofibers.

1.1.2 Electrospinning Process and Principles

The basic setup for electrospinning is shown in Fig. 1.1. The device consists of a container for carrying polymer solution or melt, a high voltage power supply, a capillary tube (needle or pipette tip) spinneret, and a collector. In a solution-based electrospinning process, a polymer solution is delivered to the tip of the nozzle. If no electric field is applied, the solution droplet from the tip will fall off under gravity. However, once a high voltage is applied to the polymer solution, a high voltage difference will be formed between the solution at nozzle tip and the grounded collector. This voltage difference can cause the deformation of droplet into a cone-shape protrusion (Taylor cone). With the increase in the applied voltage, jet initiation happens from Taylor cone immediately after the electrostatic force exceeds the surface tension of the polymer solution. Polymer jet will be ejected into the electric field. After the initial stage, the jet becomes unstable and undergoes a bending (or whipping) movement under columbic repulsion, which stretches itself thinner. With the evaporation of solvent from the filaments, dry or semidry nanofibers are formed and deposited on the grounded electrode.

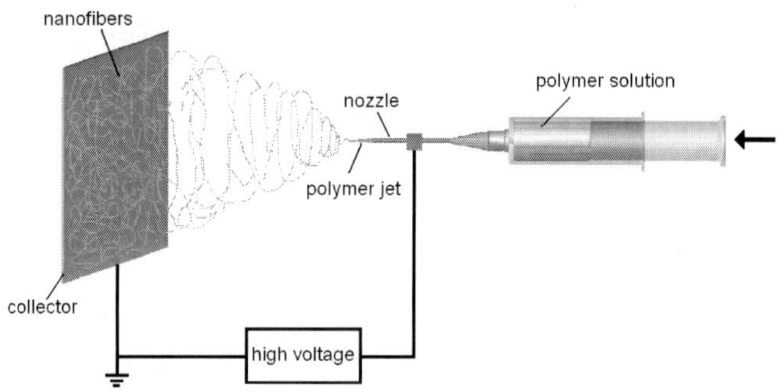

Figure 1.1 A basic electrospinning apparatus.

Fiber stretching during the electrospinning process can be divided into three stages: jet initiation, whipping (or bending) instability and fiber deposition [25].

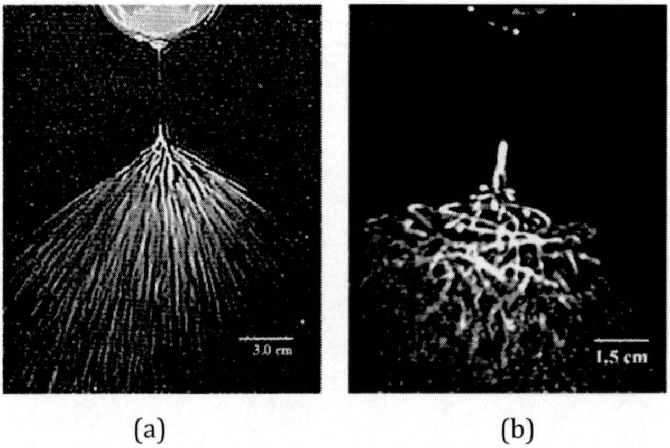

Figure 1.2 Images of an electrospinning process taken by (a) a long camera exposure time and (b) a high speed camera [26].

Fiber stretching is a very quick process that takes place in milliseconds from the initial jet to dry fibers [26, 27]. Early observation on the electrospinning process was confined by camera limitations. In a long exposure time, the jet appeared to be split into many smaller jets (Fig. 1.2a), so the thinning of fiber in electrospinning was mistakenly attributed to the repeated jet splitting. Later on, observations based on a high-speed video camera [26] indicated that a single jet followed a bending, winding, spiraling, and looping path after the initial stable stage (Fig. 1.2b). The jet became longer and thinner during this instability process. Theoretical models have been proposed to describe the bending instability [26, 28–31]. The most important contributions in this direction were made by Reneker and Yarin [26, 28]. In their research, a viscoelastic model for the electrified jet was developed and the net normal force, F, acting on a jet element was given by Eq. 1.1

$$dF = |k| n d\zeta \left[\pi a \sigma - e^2 \ln \frac{L}{a} \right] \tag{1.1}$$

where k is the jet curvature, ζ is the co-ordinate system along the central axis of the bend jet, a is the jet cross-section radius, σ is the surface tension coefficient, e is the charge density, and L represents the jet length scale.

The momentless quasi-one-dimensional equation was given using a Lagrangian parameter "frozen" into the jet elements:

$$\lambda f = \lambda_0 f_0 \tag{1.2}$$

$$\begin{aligned} \rho \lambda_0 f_0 \frac{\partial V}{\partial t} &= \tau \frac{\partial P}{\partial s} + \lambda |k| P n - \rho g \lambda_0 f_0 k \\ &+ \lambda |k| \left(\pi a \sigma - q_{el} \ln \frac{L}{a} \right) n - \lambda e \frac{U_0}{h} k \end{aligned} \tag{1.3}$$

Equation 1.2 is the continuity equation with λ being the geometrical stretching ratio, so that $\lambda ds = d\xi$, and $f = \pi a^2$ the cross-sectional area. Subscript zero denotes the parameter values at time $t = 0$. Equation 1.3 is the momentum balance equation in which λ represents the geometrical stretching ratio, $f = \pi a^2$ is the cross-sectional area, ρ is the liquid density, V is velocity vector, P is the longitudinal force in the jet cross-section, $\frac{U_0}{h}$ is the outer electric field strength, and q_{el} is the net coulomb force acting on a jet element from all the other elements depending on e and the current overall configuration of the jet. The right-hand side of Eq. 1.3 includes the longitudinal internal force of rheological origin acting on the jet (the first two terms), the gravity force (the third term), the bending electrical force and the stabilizing effect of the surface tension (the fourth term), and the electrical force acting on the jet from electric field created by the potential difference of the starting point of the jet and the collector.

The whipping instability actually arises from two opposite actions: surface tension and charge repulsion. The surface tension tends to destabilize the jet by minimizing jet surface, while the charge repulsion tends to stabilize the jet in filament shape. When charge repulsion dominates, the perturbation to the jet centerline grows, and the jet becomes bent. As the jet thins away from the spinneret, the charge density grows until the charge repulsion overcomes the surface tension, after which the jet whipping begins.

The whipping instability results in stretching the jet thinner. During the whipping process, the jet is stretched under the action of axisymmetric and non-axisymmetric instabilities due to the perturbations of surface charges. The axisymmetric instability comes from the perturbations of the surface charges along the jet axis. Different sections are affected by different strengths of force that results in an uneven jet. However, the non-axisymmetric instability comes from the perturbations of surface charges around the circumference of the jet. It leads to a localized torque and accounts for the whipping motion of the jet. Under the action of the non-axisymmetric instability, the jet tends to be uniform [32]. Figure 1.3 illustrates that a conical envelope contains only a single rapidly whipping jet. The whipping theory is further verified by the successful production of side-by-side and core-sheath electrospun nanofibers [33, 34].

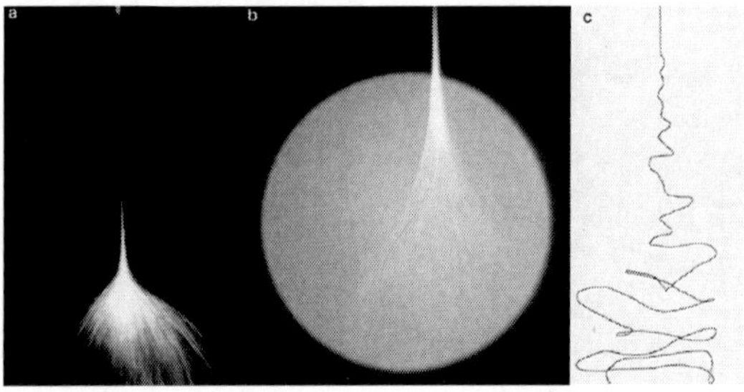

Figure 1.3 (a) Instability region in an electrified polyethylene oxide (PEO)-water jet; (b) instability region in an electrified PEO-water jet, superposition of the whipping jet (exposure time = 18 ns), and the envelope (exposure time = 1/250 s), the whipping jet can be seen as a dark line; diameter of circle = 15 mm; (c) instability region in an electrified PEO-water jet, exposure time = 18 ns and a trace of the jet is shown, vertical distance = 20 cm [35].

Viscous polymer solutions are the main liquids being studied in electrospinning, and almost all polymers that are soluble in appropriate solvents have been electrospun into nanofibers. Table 1.1

summarizes the thermoplastic polymers used for electrospinning of nanofibers. These polymer systems include synthetic polymers and natural macromolecules.

Table 1.1 Electrospun nanofibers from thermoplastic materials

Polymers	Solvents	Authors/Ref
CA	DMAC	Liu and Hsieh [36]
PVA	Water	Koski et al. [37]
Poly (2-hydroxyethyl methacrylate)	DMF	Gupta et al. [38]
Poly(2-hydroxyethyl methacrylate)	Monomers	Kim et al. [78]
PEI	1-Methyl-2-pyrrolidinone	Moon et al. [39]
PEI	TCE	Choi et al. [79]
PEO	Ethanol/water	Theron et al. [40]
PET	TFA/dichloromethane	Veleirinho et al. [41]
PET	TFA	Ma et al. [63]
Polyurethane elastomer	THF/ DMF	Lee et al. [42]
PLLA	Dichloromethane	Inai et al. [43]
PC	THF /DMF	Shawon and Sung [44]
PVCz	Dichloromethane	Bognitzki et al. [45]
PVCz	TCE	Lyoo et al. [75]
PS	THF	Casper et al. [46]
PBI	DMAC	Kim et al. [47]
Nylon 6	HFIP/ Dichloromethane	Fong et al. [48]
PVC	THF/DMF	Lee et al. [49]
PCL	Acetone	
SBS triblock copolymer	75% THF/25% DMF	Fong et al. [50]
PMMA-r-TAN	Mixed solvent of toluene and dichloromethane	Deitzel et al. [51]
PPTA	95–98 wt% sulphuric acid	Srinivasan et al. [14]

(*Continued*)

Table 1.1 (*Continued*)

Polymers	Solvents	Authors/Ref
Collagen	HFIP	Matthews et al. [52]
Fibrinogen	HFIP	McManus et al. [53]
Silk fibroin	Formic acid	Min et al. [54]
Chitin	HFIP	Noh et al. [55]
Chitosan	Formic acid/TFA/acetic acid	Ohkawa et al. [56]
PVP	Dichloromethane	Munir et al. [57]
PGA	HFIP	Boland et al. [58]
PHB	Chloroform or dichloromethane	Hsu [59]
PEU	DMAC	Caracciolo et al. [60]
PAN	Dichloromethane	Gu et al. [61]
PI	DMAC	Chen et al. [62]
PI	Formic acid	Mit-uppatham et al. [76]
PTT	TFA/Methylene chloride	Khil et al. [64]
PVDF	DMAC	Choi et al. [65]
PFDS	THF /DMF (9:1)	Wang et al. [66]
PP	Cyclohexane, acetone, and DMF (80/10/10 by weight ratio)	Lee et al. [67]
POM	HFIP	Kongkhlang et al. [68]
PPy	DMF	Chronakis et al. [69]
PANi	Chloroform	[70]
Levan	Distilled water	Manandhar et al. [71]
PLA	Chloroform	Zhao et al. [72]
PEVA	Dichloromethane	Sanders et al. [73]
PMMA	Acetone	Piperno et al. [74]
PLGA	THF	Xin et al. [77]

Polymers	Solvents	Authors/Ref
Poly (ferrocenyldimeth-ylsilane)	Mixture of 90 wt% THF and 10 wt% DMF	Chen et al. [80]
EVOH	Alcohol/water	Kenawy et al. [81]
Polymetha-phenylene isophthalamide	BMIM (BF_4)	Yang et al. [82]

Abbreviations: BMIM, 1-butyl-3-methyl imidazolium tetrafluoroborate; CA, cellulose acetate; DMAC, *N,N*-dimethylacetamide; DMF, *N,N*-dimethylformamide; EVOH, poly(ethylene-co-vinyl alcohol); HFIP, 1,1,1,3,3,3-hexa fluoro-2-propanol; PAN, polyacrylonitrile; PANi, polyaniline; PBI, polybenzimidazole; PC, polycarbonate; PCL, polycaprolactone; PEI, Poly (etherimide); PEO, polyethylene oxide; PET, polyethylene terephthalate; PEU, poly(ester urethane); PEVA, polyethylene-co-vinyl acetate; PFDS, polyferrocenyldimethylsilane; PGA, poly(glycolic acid); PHB, polyhydroxybutyrate; PI, polyimide; PLA, poly lactic acid; PLGA, poly(d,l-lactide-*co*-glycolide); PLLA, poly-L-lactide; PMMA, polymethacrylate; PMMA-r-TAN, poly (methyl methacrylaterandom); POM, polyoxymethylene; PP, polypropylene; PPTA, poly (p-phenylene terephthalamide); PPy, polypyrrole; PS, Polystyrene; PTT, poly (trimethylene terephthalate); PVA, polyvinyl alcohol; PVC, polyvinyl chloride; PVCz, polyvinylcarbazole; PVDF, poly(vinylidene fluoride); PVP, polyvinylpyrrolidone; PS, polystyrene; SBS, styrene-butadiene-styrene; TCE, 1,1,2,2-tetrachloroethane; TFA, trifluoroacetic acid; THF, tetrahydrofuran.

Besides these polymers, ceramic, metal, and organic–inorganic hybrid materials have been used for electrospinning as well. Inorganic nanofibers are normally electrospun using sol-gel-based solutions. Sols are dispersions of colloidal particles in a liquid, and gels are interconnected rigid inorganic networks with pores of sub-micrometer dimensions [83]. Sol-gels are typically prepared by hydrolysis and condensation of precursors. Sol-gel solutions for electrospinning are generally prepared by hydrolysis of metal alkoxide in an acidic condition, because the hydrolysis in acidic medium is more likely to form spinnable sol solutions. Despite this, normally prepared sol solution is still difficult to form fibrous material in electrospinning. To improve the spinnability, sol-gel solution is usually blended with a polymer for improving viscosity and fiber-forming property. Because of this, a high temperature is needed to remove the organic components from the electrospun nanofibers. The calcination process generally leads to a rough surface, porous fiber structure, and reduction in fiber diameter.

To improve the fiber properties and add new functionalities to nanofibers, nano-sized filler materials, such as clay, carbon nanotubes, and metal nanoparticles, have been added to the polymer solutions for electrospinning.

1.1.3 Fiber Morphology

Nanofibers prepared by electrospinning can have different morphologies, such as uniform nanofibers with round cross-sections, beads-on-string structures, or individual beads. Factors that influence the fiber morphologies include polymer concentration, solution viscosity, solution conductivity, solution surface tension, and solvent properties. Also, operating parameters such as applied voltage, collecting distance (also as nozzle-to-collector distance), and flow rate of polymer solution can change the fiber morphologies.

Polymer concentration is the most important factor affecting an electrospinning process, because it also affects many other solution properties such as viscosity, conductivity, and surface tension. For a polymer with a reasonable molecular weight, changing polymer concentration for electrospinning can result in three different fiber morphologies, such as beads, beads-on-string, and uniform fibers (Fig. 1.4). Higher polymer concentration tends to result in coarser fibers. For the same polymer solution system, a higher viscosity suggests stronger entanglement between the polymer chains, which makes stretching more difficult. For some polymers, when the polymer concentration is increased, the charge density in solution also increases. This will lead to a stronger force to stretch the filaments. Hence with increased polymer concentration, the fibers are stretched under two opposing forces.

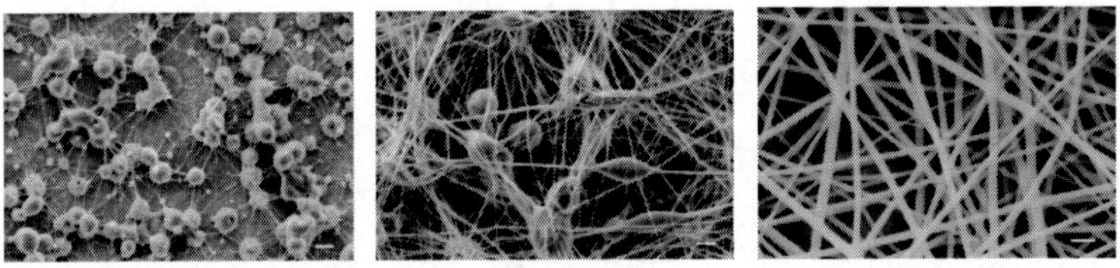

(a) (b) (c)

Figure 1.4 Electrospun products from poly(vinylidene fluoride-co-hexafluoropropylene) (PVdF-HFP) solutions (in DMF/THF 1:1 vol/vol) with different concentrations (a) individual beads from 3 wt% solution, (b) beads-on-string structures from 7% solution, and (c) uniform fibers from 11 wt% solution (scale bar: 1 μm).

An absolutely non-conductive solution cannot be electrospun because no electric charges could be transferred to the droplet surface. However, if the solution is highly conductive, extra charge could not be accumulated on the droplet surface. As a result, fluid could not be drawn into fine filaments because of the very small stretching force. The solution for electrospinning should be in the semi-conducting range.

For some polymers, beaded fibers are always collected even when the concentration is very high. In this case, increasing the solution conductivity via the addition of other chemicals to the solution is found to be an effective approach to stop the formation of beaded fibers. Additives such as inorganic salts, surfactants, and polymers have been added into polymer solutions to increase conductivity. With the increase in solution conductivity, uniform nanofibers can be obtained, and the fiber diameter is even reduced with the increase in additive concentration. Some of these additives can also decrease the solution surface tension, which also facilitates the electrospinning process.

Solvent in solutions also affects fiber morphologies. Dry fibers are collected only when the solvent evaporates completely from the filaments, before they deposit on the collector. For some solvents of low volatility, wet fibers would be collected, which merge together to form an interconnected web structure [84]. For some highly volatile solvents, rapid evaporation of the solvent cools the filament and moisture can be frozen on the filament surface, resulting in the formation of nano/micro-sized pores on the fiber surface [85].

Applied voltage is the source of stretching force in electrospinning, and it is the primary factor influencing the electrospinning process. Since the bending instability and jet stretching become more intensive under higher applied voltage, finer fibers could be prepared when the applied voltage is increased. Because of the stronger bending instability under a higher applied voltage, fiber morphology can be changed from uniform fibers to beads-on-string structure. However, the applied voltage cannot be increased unlimitedly. For a needle electrospinning setup, when the applied voltage is increased to about 30 kV, "corona discharge" occurs and stops the electrospinning process.

Electrospinning process typically uses direct-current (DC) voltage, while alternating-current (AC) voltage can also drive the electrospinning process. Since AC voltage is not as safe as DC voltage, especially at a high voltage value, only a few papers reported on the use of AC power supply for electrospinning experiments [86]. It was reported that aligned nanofibers could be easily collected on a rotating mandrel when AC power supply was employed, and the reason for this was that both positive and negative charges were carried onto the collector, resulting in neutralizing and minimizing the accumulation of surface charges.

During electrospinning, the evaporation of solvent from filament takes time. If the collecting distance is short, wet fibers could be collected; therefore, a solvent of higher volatility should be selected [84, 87]. Alternatively, extra gas flow or higher operating temperature can be used to speed up the evaporation process. A longer collecting distance will give the solvent more time to evaporate. If the electric field intensity is kept the same, longer collecting distance normally leads to thinner fibers [88] because of the longer stretching time during electrospinning. However the electric field strength will decrease if the applied voltage remains the same.

The flow rate for feeding the polymer solution during electrospinning should match the rate of exhausting the polymer solution from the nozzle tip, and coarse or beaded fibers are always resulted from a high flow rate [89].

Apart from the abovementioned factors, some other parameters such as environment temperature, humidity, extra gas flow, solvent polarity can also influence the electrospinning process and fiber morphology.

1.2 Characteristics of Electrospun Nanofibers

Electrospun nanofibers exhibit many unique characteristics [90], such as

- High surface-to-mass (or volume) ratio;
- High porosity with excellent pore interconnectivity;
- Flexibility with reasonable structural mechanical strength;
- Low unit weight;
- Extensive selection of the polymer materials;
- Ability to integrate other materials (e.g., chemicals, polymers, biomaterials, and nanoparticles) into nanofibers through electrospinning;
- Ability to adjust fiber dimension, alignment, chemical composition, and other fiber properties (such as strength, weight, elasticity, porosity, charged surface area);
- Ability to control secondary structures of nanofibers in order to prepare nanofibers with core/sheath structures, side-by-

side structure, hollow nanofibers, and nanofibers with porous structures.

These unique properties plus the functionalities from polymer materials make electrospun nanofibers very useful for many advanced applications. Details about the applications of electrospun nanofibers are introduced in Chapter 2.

1.3 Improved Electrospinning Techniques

Since the emergence of electrospinning technology, many innovative electrospinning systems have been developed. Basically, according to the spinneret and spinning mode, electrospinning can be broadly classified into needle electrospinning and needleless electrospinning. In needle electrospinning, a needle-like spinneret is used, and each needle produces only one fiber or filament each time. At the needle tip, a sharp "cone shape" is formed under the electric field. For needleless electrospinning, multiple cones are produced from a surface not containing any needle or tiny open structure.

Inspired by the core-shell, hollow, and side-by-side bicomponent fibers, nanofibers with such bicomponent cross-sectional configurations can also be produced by electrospinning using special spinnerets, and the bicomponent nanofibers can be classified into two main types: core-sheath and side-by-side bicomponent nanofibers.

The core-sheath nanofibers are prepared using a co-axial spinneret [91, 92]. In core-sheath electrospinning, two polymer solutions are fed separately via a coaxial spinneret to the spinneret tip (Fig. 1.5a). When a mineral oil is used as the core fluid, hollow nanofibers can be simply produced from the core-sheath electrospinning (Fig. 1.5b).

Core-sheath electrospinning is similar to conventional needle electrospinning, except that two polymer fluids are co-electrospun simultaneously from one spinneret. In core-sheath electrospinning, the polymer solutions at the spinneret tip are mainly charged on the sheath surface. Upon the formation of polymer jet, the extrusion of the polymer sheath layer also causes the core solution to deform into a conical shape due to the contact friction. While the sheath solution is stretched, great interface elasticity between the core and sheath fluids stabilizes the core fluid. A compound co-axial jet is thus developed [93]. As long as this compound jet is stable, the core component will be uniformly incorporated into the sheath component to form core-sheath fibers.

Reznik et al. [94] studied the positions of inner needle and outer needle in core-sheath electrospinning. They found that the formation of core-sheath jet and nanofibers was greatly facilitated when the core spinneret protrudes outside the sheath spinneret by around half of its radius size. Otherwise unsuccessful core-sheath electrospinning could result. The operating conditions are also very important for core-sheath electrospinning. At a low applied voltage, intermittent polymer jet and dripping occur. At a higher applied voltage, stable "Taylor cone" and continuous electrospinning result. Much higher applied voltage leads to the formation of multiple jets from single spinneret [93].

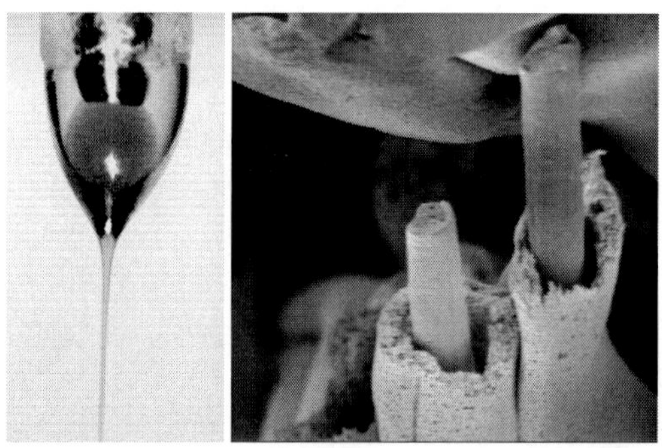

(a)

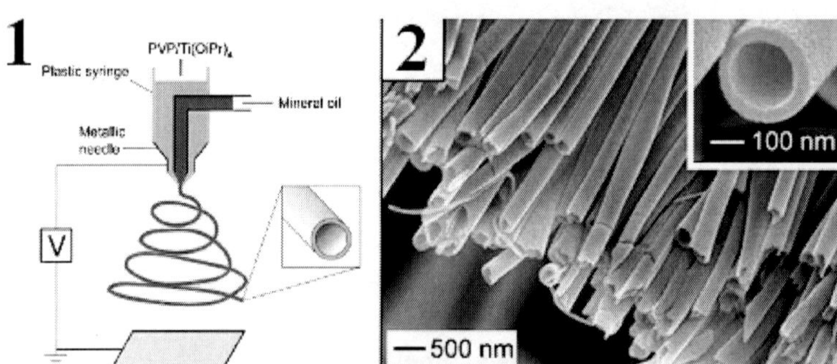

(b)

Figure 1.5 (a) Core-sheath electrospinning and a SEM image of core-sheath nanofibers [91], (b) modified core-sheath electrospinning process to produce hollow nanofibers (1—electrospinning setup; 2—electrospun hollow nanofibers) [92].

Side-by-side electrospinning nanofibers were first reported by Gupta and Wilkes [95], who used two syringes to make two polymer solutions to contact at the spinneret tip for electrospinning (Fig. 1.6a). However, this work did not prove the existence of "side-by-side" bicomponent fiber structure. Later Lin et al. [33] demonstrated that when a microfluidic device was used as a spinneret to put two polymer solutions side-by-side into the same microfluidic channel for electrospinning, nanofibers having a side-by-side cross-sectional configuration can be produced successfully (Fig. 1.6b). The formation of side-by-side bicomponent nanofiber structure was clearly proven by selective removal of one polymer component from the fiber, and the remaining fibers showed a "U"-shape cross-section (Fig. 1.6c).

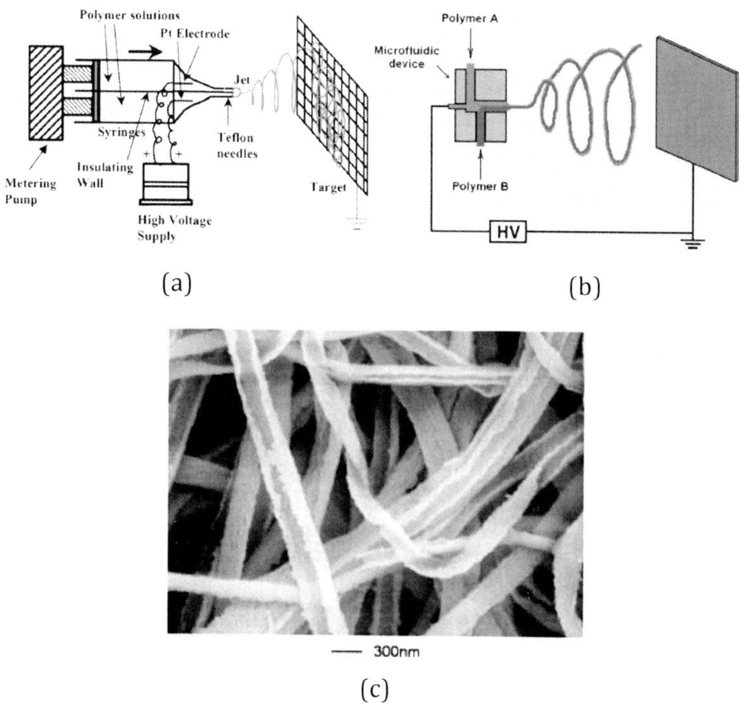

Figure 1.6 (a) Side-by-side electrospinning [95], (b) side-by-side electrospinning setup using a microfluidic device [33], (c) a SEM image of side-by-side bicomponent nanofibers with one fiber side removed [33].

Bhaskar and Lahann [96] further improved the side-by-side spinneret and used it to produce multi-component fibers (Fig. 1.7).

However, the as-prepared fibers were on the micro scale rather than nano scale.

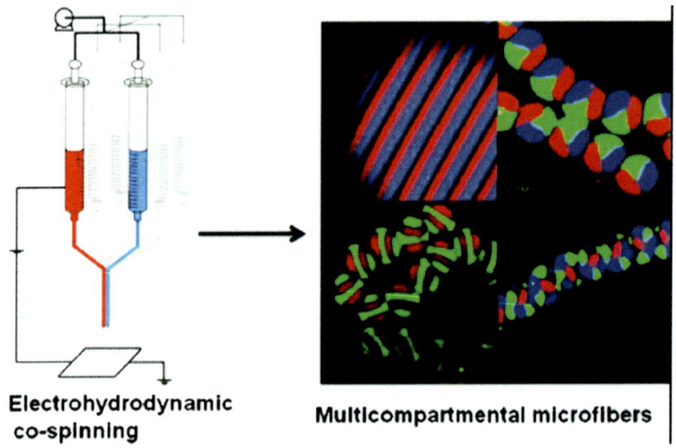

Figure 1.7 Schematics of a multi-compartmental electrospinning setup and fibers [96].

In conventional electrospinning, DC voltage is usually applied. It has been reported that multi-stranded threads are fabricated using AC electrospinning (Fig. 1.8) [97]. The process was obviously different from the DC one, as a visible thread was seen emerging downstream from the needle instead of the mesh-like morphology observed in DC spinning. This thread could be more than a meter long and did not have any significant attraction toward the ground electrode, and consequently it could be easily deflected away and collected [97].

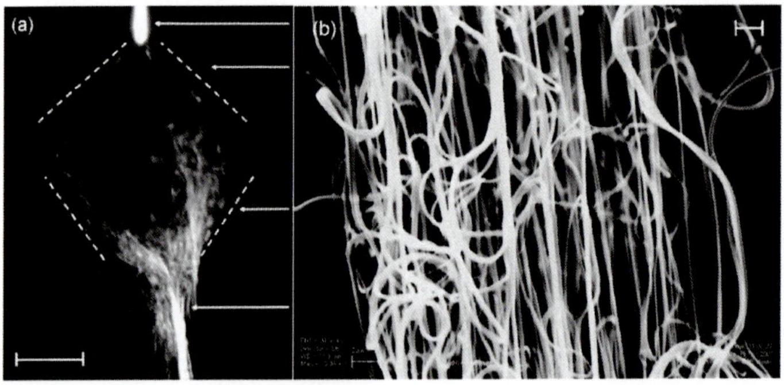

Figure 1.8 (a) AC electrospinning and (b) a SEM image of as-spun nanofibers (scale bar: 2 microns) [97].

Melt electrospinning of fibers does not require the use of solvents and hence is environmentally friendly. Many polymers have been successfully electrospun into ultrafine fibers from their melts, including polypropylene and poly(ethylene terephthalate) [98, 99], polyalirate [99], low-density polyethylene [100], and poly-(ethylene glycol-block-ε-caprolactone) [101]. Figure 1.9 shows two melt-electrospinning setups and the electrospun nanofibers. It is worth noting that in melt electrospinning, the high voltage power supply is sometimes connected to the collector to avoid the interference between the high voltage and electric heating device in the spinneret [100].

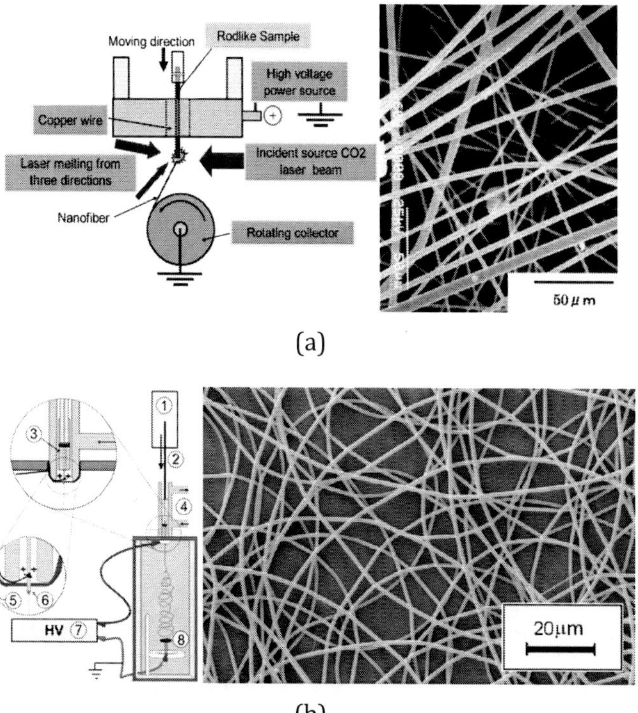

Figure 1.9 Schematic illustration of two melt-electrospinning setups and the SEM images of electrospun fibers, (a) [99], (b) [101].

1.4 Fiber-Collecting Modes

In conventional electrospinning, a grounded metal plate is typically used as the nanofiber collector, which enables nanofibers to deposit

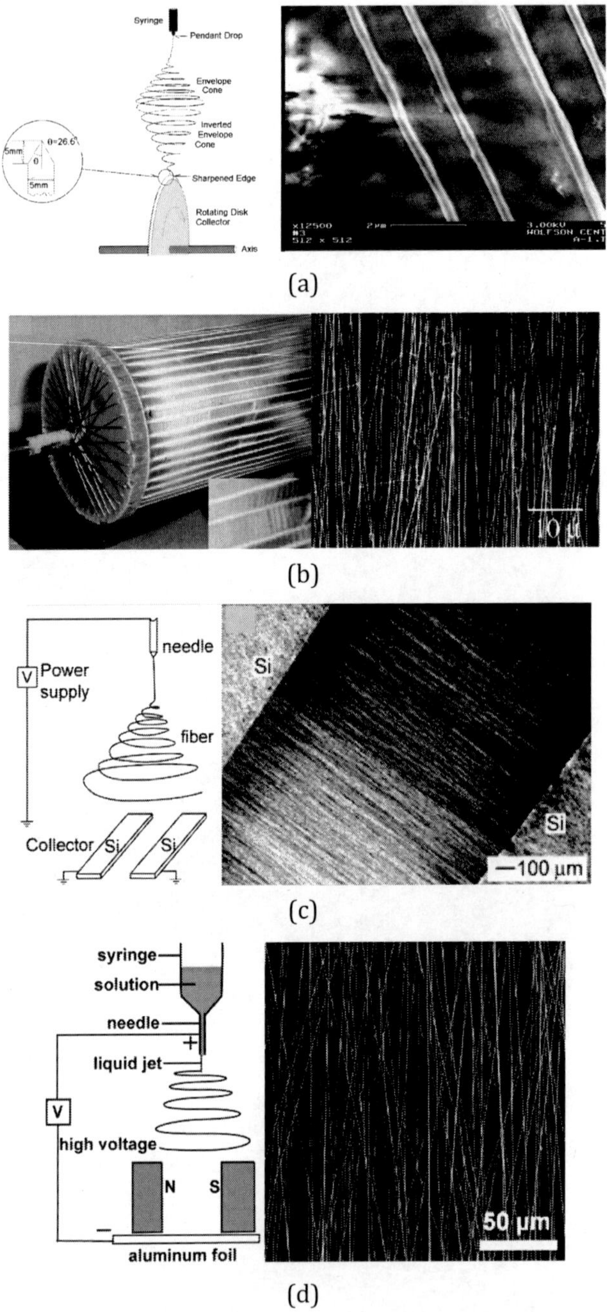

Figure 1.10 Setup for electrospinning aligned nanofibers and obtained nanofibers, (a) [102], (b) [103], (c) [105], (d) [106].

on the plate randomly. Aligned nanofibers have been collected directly from an electrospinning process using a disc collector (Fig. 1.10a) [102]. However, the disc collector is only able to collect aligned nanofibers in small amount. Katta et al. [103] used a rotating drum consisting of many parallel metal wires to collect aligned nanofibers (Fig. 1.10b). With these rotating drum collectors, well-aligned nanofibers can be obtained initially; the alignment decreases after prolonged fiber collection. Aligned nanofibers were also collected using motionless parallel electrodes collector (Fig. 1.10c) [104, 105] or magnetic collector (Fig. 1.10d) [106]. When a magnetic field was used, magnetic nanoparticles were needed to be added into polymer solution, so that the fiber was responsive to the magnetic field.

Carnell et al. [107] collected nanofibers with a precise control. Their work (Fig. 2.11) applied counter voltage with a electrode to influence the fiber placement and alignment by controlling the shape and magnitude of the electric field, allowing controlled fabrication of highly aligned electrospun fibers and mats [107].

Aligned pseudo-woven mats with reasonably controlled porosity can be obtained with this method, and fiber placement and alignment on both the microscale and nanoscale can be archived.

Figure 1.11 Collection with auxiliary counter electrode and SEM image [107].

Hellmann et al. [108] reported a technique to precisely control the deposition of nanofibers by reducing the spinning distance from 10–50 cm to the millimeter range and applying a counter electrode/substrate, which can be subjected to precise motions on the deposition plane (Fig. 1.12).

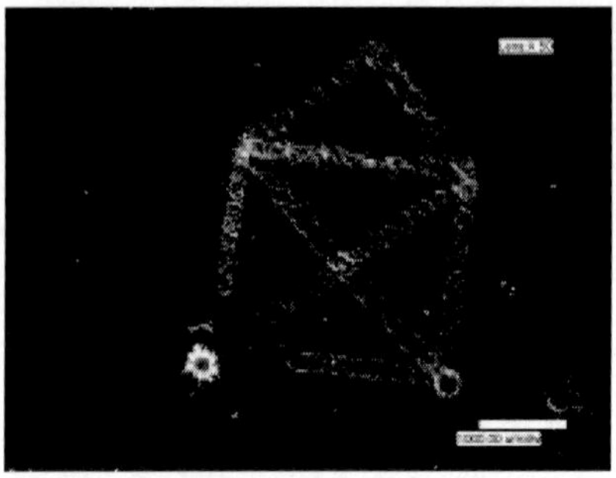

Figure 1.12 "House of Santa Claus" motif written by PAN nanofibers (distance = 2.2 mm, voltage = 0.47 kV) [108].

Zhang and Chang [109] have reported a static method to collect three-dimensional (3D) fibrous tubes for biomedical and industrial applications. By designing 3D collectors, fibrous tubes with different macroscopic configurations (length, diameter, and shape) can be fabricated, and multiple tubes with various interconnected tubular structures could be fabricated using removable collecting templates (Fig. 1.13).

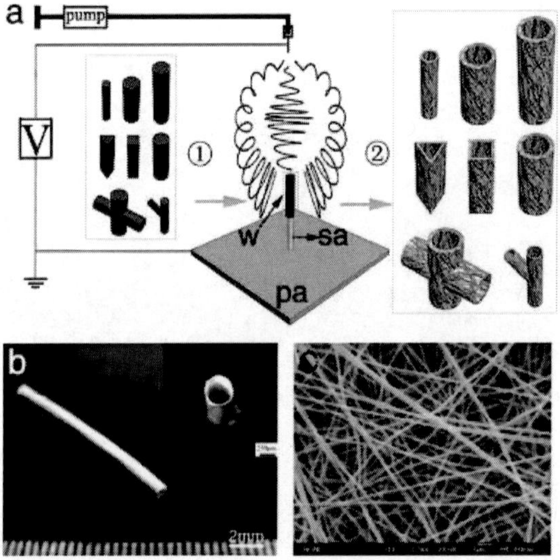

Figure 1.13 (a) Schematic illustration of fibrous tubes electrospun using 3D columnar collectors. (①: 3D columnar collectors and ②: fibrous tubes; *w*, working collector; *pa*, plane assistant collector; *sa*, stick assistant collector); (b) fibrous tube electrospun (diameter, 500 μm, inset is the cross-section image); (c) a SEM image of fibers assembled in the fibrous tube [109].

Yarns are fundamental elements to construct a fibrous architecture such as woven fabrics. To broaden the applications of electrospun nanofibers, collection of nanofibers into yarns is essential. Smit et al. [110] described a simple technique to collect nanofibers into continuous fiber bundles using a water reservoir as collector and continuously drawing the deposited fiber web off the reservoir bath (Fig. 1.14).

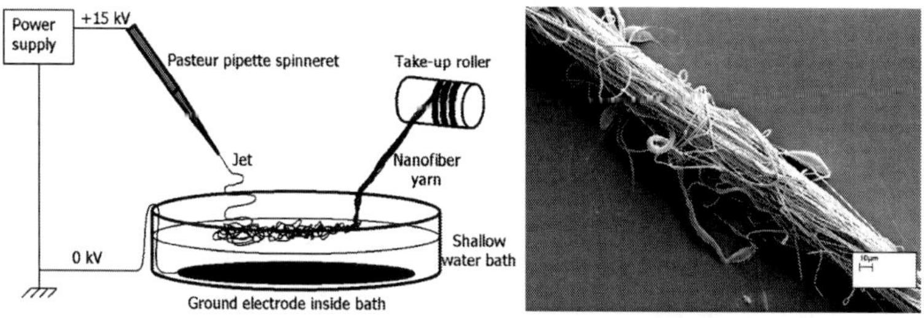

Figure 1.14 An electrospinning setup using a water reservoir as collector to produce nanofiber yarns and a SEM image of the yarn collected [110].

Dalton et al. [111] used two metal discs to collect aligned nanofibers between the discs. When one of the discs rotates during the fiber collection, twisted short nanofiber yarn can be obtained (Fig. 1.15).

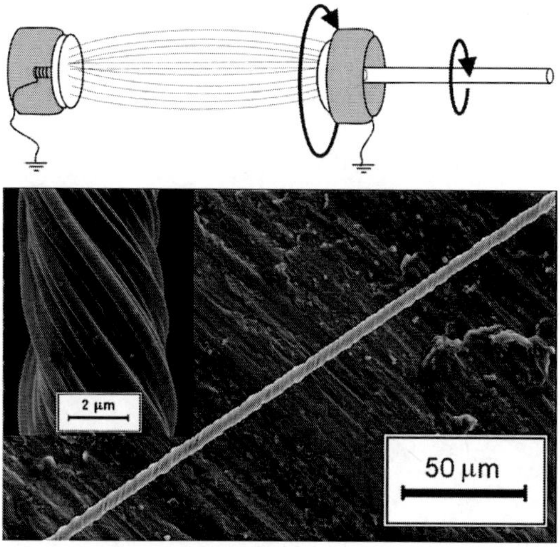

Figure 1.15 Dual disc fiber collector and a SEM image of the twisted nanofiber yarn [111].

Pan et al. [112] used two oppositely charged metallic needles that were connected to positive and negative voltages to produce aligned electrospun fibers. As shown in Fig. 1.16, the fibers coming out of the two needles combine to form a fiber bundle, which is wound by a high-speed cylinder collector.

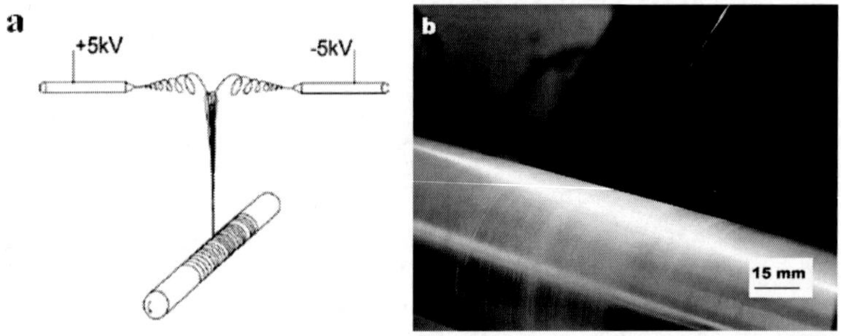

Figure 1.16 A schematic electrospinning setup (a) and collected nanofiber yarn (b) [112].

Chapter 2

Emerging Applications of Electrospun Nanofibers

Although many reviews on nanofibers have been published [25, 113–118], new applications are found continuously. In this chapter, we expand on an earlier review on nanofiber applications [116] and summarize the recent research advances in the application of electrospun nanofibers. The state-of-the-art applications in functional areas, such as biomedical, energy harvest and storage, and environmental protection are presented in detail.

2.1 Biomedical

2.1.1 Tissue Engineering Scaffolds

Tissue engineering is an emerging interdisciplinary and multidisciplinary research field which involves the use of living cells, manipulated through their extracellular environment or genetically to develop biological substitutes for implantation into the body, and/or to foster remodeling of tissues in some active manner. A basic principle of tissue engineering is illustrated in Fig. 2.1. The purpose of tissue engineering is to repair, replace, maintain, or enhance the function of a particular tissue or organ.

Needleless Electrospinning of Nanofibers: Technology and Applications
Tong Lin and Xungai Wang

ISBN 978-981-4316-35-4 (Hardcover), 978-981-4316-84-2 (eBook)
www.panstanford.com

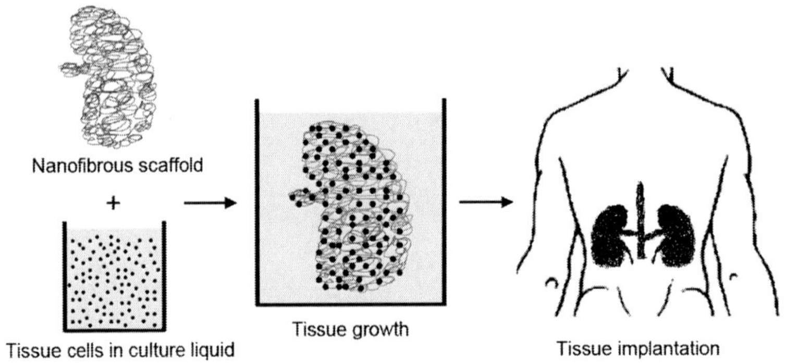

Figure 2.1 Illustration of tissue engineering [116].

The core technologies intrinsic to this effort can be organized into three areas: cell technology, scaffold construct technology, and technologies for in vivo integration. The scaffold construct technology focuses on designing, manufacturing, and characterizing three-dimensional scaffolds for cell seeding and in vitro or in vivo culturing.

For a functional scaffold, a few basic requirements have to be met [119]. First, a scaffold should possess a high degree of porosity, with an appropriate pore size distribution. Second, a large surface area is needed. Third, biodegradability is often required, with the degradation rate matching the rate of neo-tissue formation. Fourth, the scaffold must possess the required structural integrity to prevent the pores of the scaffold from collapsing during neo-tissue formation, with the appropriate mechanical properties. Finally the scaffold should be non-toxic to cells and biocompatible, positively interacting with the cells to promote cell adhesion, proliferation, migration, and differentiated cell function.

Among all biomedical materials under evaluation, electrospun nanofibrous scaffolds have exhibited great performance in cell attachment, proliferation, and penetration, through both in vitro and in vivo trials. They have potential for application in many aspects of tissue engineering.

2.1.2 Blood Vessels

Blood vessels vary in sizes, mechanical and biochemical properties, cellar content, and ultra-structural organization, depending on their location and specific function. It is required that the vascular grafts engineered should have desired characteristics. Blood vessel replacement, particularly a fine blood vessel (diameter < 6 mm), has remained a great challenge.

Culturing vascular endothelial cells (ECs) on electrospun nanofiber mats has been investigated. A comparison of cell growth on electrospun poly(L-lactide-co-ε-caprolactone) (P(LLA-CL)) nanofibers and a smooth P(LLA-CL) cast film revealed that the function of vascular ECs on the P(LLA-CL) cast film, rather than on the electrospun nanofiber mat, was enhanced [120]. But because the electrospun nanofiber mats can give good support during the initial growth of vascular smooth muscle cells [121], smooth film combining with electrospun nanofiber mat could form a good 3D scaffold for blood vessel tissue engineering.

The effect of fiber diameter on ECs culturing was examined using electrospun CA fibers with three different diameter ranges: 0.01–0.2, 0.2–1, and 2–5 μm [122]. It was interesting to note that the ECs showed a growth preference towards larger fibers. Similar results were also found on PCL nanofibers with five different diameters [123]; cell penetration increased with the increased fiber diameter, and unobstructed cell delivery was observed only on the largest fibers (12.1 μm in diameter).

To mimic the morphological and mechanical characteristics of a native blood vessel scaffold, bilayered electrospun nanofiber architectures were fabricated. A scaffold consisting of a stiff and oriented PLA outer nanofiber layer and a pliable and randomly oriented PCL inner nanofiber layer was electrospun. Such a hierarchical scaffold was reported to be able to support the attachment, spread, and growth of mouse fibroblasts and human myofibroblasts [124]. A composite scaffold incorporating a highly porous poly(ester-urethane)urea (PEUU) inner layer and an external electrospun nanofiber layer was prepared for small diameter vascular grafts (Fig. 2.2a) [125]. The mechanical properties of the bilayered scaffold were comparable with those of native vessels, and the combination of the two layers enabled better cell integration and growth. Because of the porosity difference in the two layers, high cellular density was found in the inner layer (Fig. 2.2b).

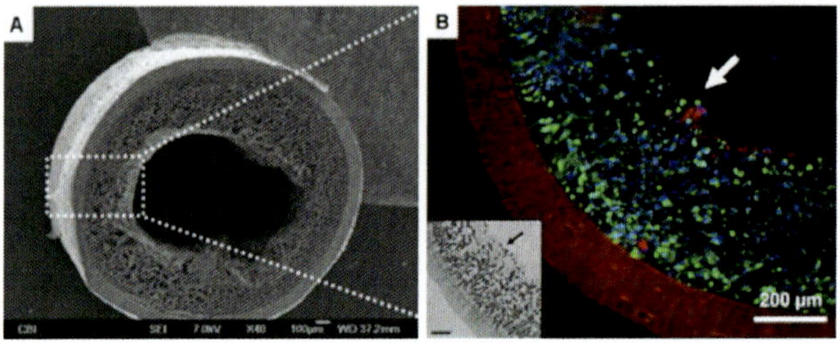

Figure 2.2 The cross-sectional images of a double-layered small diameter PEUU vascular graft before (a) and (b) after cell culturing (blue, nuclei; green, F-action; red, scaffold) [125].

Surface modification on electrospun nanofibers with natural proteins, such as collagen or gelatin, was found to be an effective way to promote ECs spreading and proliferation. Examples can be found for collagen-grafted PCL [126] and gelatin-grafted PET nanofibers [63]. The modified-nanofiber mats showed apparent enhancement in the spreading and proliferation of ECs compared with the non-grafted ones. ECs readily oriented along the aligned collagen-grafted PCL nanofibers but not on the nonwoven counterparts. On a tubular electrospun silk fibroin scaffold, the scaffold had better cell proliferation and alignment under dynamic flow conditions than static culture controls.

Since a fibroblast growth factor (FGF-2) modulates cell growth, differentiation, migration, and survival, the ability to bind FGF-2 to electrospun matrix was improved effectively by coating a bioactive recombinant fragment of perlecan on the surface of electrospun nanofibers [127]. Also, an adhesive protein, fibronectin (Fn), was grafted onto electrospun P(LLA-CL) nanofibrous scaffold, and the Fn-grafted nanofiber scaffold was observed to promote epithelium regeneration [128].

To match the mechanical characteristic of natural blood vessel, which has a low-strain mechanical response to blood flow and prevents pulsatile energy from being dissipated as heat, elastin/polymer blend nanofibers were used [129]. The preliminary cell culture studies showed that cells migrated through the full thickness of the elastin-containing grafts, but failed to migrate into pure polymer nanofiber control scaffold. To mimic the ratio of collagen and elastin in native blood vessel, electrospun nanofibers of collagen/elastin/synthetic polymer (e.g., PLGA, PLLA, PCL) (45:15:40 w/w/w) blend were assessed [130]. The as-spun nanofiber mat showed no cytotoxicity and was dimensionally stable, and its mechanical properties were similar to the native blood vessels.

In vivo assessments of cylindrical electrospun nanofiber constructs from collagen, gelatin, and synthetic biodegradable polymers such as PGA, PLA, and PGA/PLA copolymer were conducted [131]. When implanted into the interstitial space of rat vastus lateralis muscle, collagen construct was rapidly and densely infiltrated by interstitial and endothelial cells, and functional blood vessels were evident within 7 days, while gelatin and synthetic nanofiber constructs were not infiltrated to any great extent and induced fibrosis. In another rabbit aortoiliac bypass model [132], electrospun PCL/collagen scaffolds supported adherence and growth of vascular cells, and endothelialized grafts resisted adherence of platelets when exposed to blood.

2.1.3 Bones

Research on engineering bone tissues using electrospun nanofibers began with the work on muscles [133]. Cell growth on PLGA nanofibers indicated that nanofiber structure positively promoted cell–matrix and cell–cell interaction.

For bone tissue engineering, nanofibers from PCL have been extensively studied [134–136]. It was observed that Mesenchymal stem cells (MSCs) penetrated into the PCL matrix accompanied with abundant extracellular matrix after 1 week of seeding, and mineralization and type I collagen occurred at 4 weeks. An in vivo experiment was also conducted by implanting a MSCs cultured PCL construct (4 weeks) in the omenta of rats [135]. After 4 weeks, the constructs maintained the size and shape of the original scaffolds and had a bone-like appearance.

In addition to the pure PCL nanofibers, nanofibers from gelatin/PCL blend [137] and PCL composite with calcium carbonate nanoparticles and hydroxyapatite (HAp) nanoparticles [138, 139] for bone scaffolds have been investigated. The addition of 50% gelatin to PCL improved both the fiber mechanical strength and surface wettability, therefore enhancing the cell attachment and growth on the scaffold [140]. Also, the cells were observed to migrate up to 114 μm inside the scaffold within 1 week of culture. In contrast, PCL nanofibers containing these inorganic nanoparticles were expected to have higher osteoblast proliferation and differentiation.

In addition, a bond-like calcium phosphate (CaP) was coated on PCL nanofiber surface. The CaP-coated nanofiber mat showed high wettability. Because of its similar structure to the natural bone, such a mineralized electrospun scaffold is expected to be a potential cell carrier in bone tissue engineering.

A combination of PCL nanofibers and microfibers was used to develop multilayer scaffolds [141, 142]. This multilayer fiber mat contained a thin nanofiber upper layer and an inner microfiber fibrous structure. As the top nanofiber layer assisted cell attachment and spread, the inner microfiber layer provided large pores for cell migration. Both cell attachment and migration were improved. Fiber diameter has shown obvious effect on cell culture. Gelatin nanofibers were electrospun with different fiber diameters of 110 and 600 nm for osteoblastic MG63 cell growth. The cells were found to have a similar attachment feature at the first 3 days and with the extended culture time they preferred the coarser fibers. In longer term experiments, the cells differentiated to a greater extent on the scaffolds made from the smaller diameter fibers [143]. Human MSCs were investigated on electrospun PCL nanofibrous and microfibrous scaffolds. The results indicated that the cells attached and spread rapidly on the nanofibrous scaffolds, and the cells were found to differentiate into the osteoblast lineage [144].

Some nanofiber scaffolds have shown excellent biocompatibility with enhanced bone regeneration, without any inflammatory reaction. When silk fibroin nanofibers contained bone morphogenetic protein-2 (BMP-2) or HAp nanoparticles, the presence of BMP-2 supported higher calcium deposition and enhanced transcript levels of bond-specific markers, while the incorporation of HAp nanoparticles into the silk fibroin nanofibers improved the bone formation [145]. The coexistence of the two materials gave the largest improvement in calcium deposition.

Bioactive glass (BG) has also been widely studied for bone regeneration. PCL/BG composite nanofibers considerably improved the differentiation of osteoblast cells as compared to pure PCL nanofibers [146], and a short BG nanofiber/PCL matrix composite scaffold has shown greater mechanical stability and bioactivity than BG power/PCL composite [147].

Mechanical stimulation has been applied during bone regeneration, and the results showed that proper stimulation can enhance fracture healing due to the anabolic osteogenic effects [148].

Mechanical stresses induced by embedded magnetic nanoparticles were applied onto PCL scaffolds to stimulate growth, proliferation, and differentiation of preosteoblast MC3T3 cells [149].

It is worth noting that a nanoporous surface on electrospun fibers (e.g., poly(ethylene oxide terephthalate)-poly(butylenes terephthalate) (PEOT/PBT) fibers) can significantly enhance the cell proliferation and spreading on the fibers [150]. In vivo studies [151, 152] have also suggested the great potential of electrospun nanofiber scaffold in skeletal tissue engineering because of adequate mechanical properties and osteocondition.

2.1.4 Cartilages

Articular cartilage, composed of primarily type-II collagen, is the connective tissue covering the ends of long bones, and serves the functions of load bearing (shock absorber) and joint lubrication. The feasibility of using electrospun collagen (type-II) nanofibers to culture chondrocyte has been examined [153, 154]. Collagen nanofibers were observed to be compatible with chondrocytes and readily infiltrated the scaffold surface and interior. PCL nanofiber mat was also used to culture chondrocyte cells in a serum-free medium. It was found that PCL matrix promoted phenotypic differentiation and cellar proliferation [155], suggesting that the unique structural features and the biomimetic nature of PCL nanofibers enable them to support the differentiation of cells without the aid of serum.

The importance of surface topography was proven via comparison of cultured chondrocytes on poly (3-hydroxybutyrate-co-3-hydroxyvalerate) (PHBV) nanofiber mat and cast PHBV film [156, 157]. Although the PHBV nanofiber mat showed a higher water contact angle than the cast PHBV film, more cells were observed to attach on the PHBV nanofiber mat with greater spread morphology. Also, chitosan/PEO nanofiber mat showed better cell adhesion than its flat film counterpart [158].

In order to improve cell attachment and growth, collagen [159] and gelatin [160] were grafted onto PLLA nanofibers after plasma treatment. The results have shown better cell differentiation than those of pristine PLLA nanofiber mat. Research has also shown that pore size in scaffold is very important to achieve effective chondrocyte cultivation, and micron-sized fibrous mat could have better MSC chondrogenesis than nano-sized fibrous scaffold [161].

Aimed at obtaining highly porous scaffolds for cartilage regeneration, several different approaches have been used. PBT nanofibers were electrospun on a novel 3D PEOT fibrous structure to form composite scaffold [162], electrospun PLA nanofibers were coated onto PCL microfibers to form a scaffold with a porosity as high as 97% [163], and a 3D scaffold was prepared by rolling up a double-layered mat containing a layer of microfibrous mat and a layer of electrospun nanofibers [164]. Better chondrocyte attachment, spreading, differentiation, and, more importantly, cell infiltration were observed on these innovative 3D scaffolds.

2.1.5 Muscles

Collagen nanofibers were first used to assess the feasibility of culturing smooth muscle cell [52]. The cell growth on the collagen nanofibers was promoted, and the cells were well integrated into the nanofiber network after 7 days of seeding. Smooth muscle cells also adhered and proliferated well on nanofiber mats electrospun from other polymers, such as PEUU/collagen blend [165], P(LLA-CL) [121], PCL/collagen [166], PS [88], and collagen/chitosan blend [167]. The incorporation of collagen into nanofibers was observed to improve fiber elasticity and tensile strength, and increase the cell adhesion.

The fiber surface wettability influences cell attachment. It was reported that when the PS nanofibers were treated with Argon plasma, the fiber wettability was increased significantly. As a result, cell attachment was increased by two folds [88]. The alignment of nanofibers can induce cell orientation and promote skeletal muscle cell morphologies and aligned myotube formation. The cells cultured on the aligned nanofibers exhibited an alignment factor of 0.74, compared with 0.19 on the randomly orientated scaffold [88].

Nanofiber mat electrospun from a commercial polyester urethane (DegraPol®) was used for skeletal muscle tissue engineering. Three different cell lines, murine myoblast cell line (C2C12), rat myoblast cell line (L6), and primary human satellite cells (HSCs), were observed to adhere to, and proliferate well on, the nanofiber mats [168]. The DegraPol® scaffolds were able to assist skeletal myogenesis by improving myoblast adhesion, myotube alignment, and noncoplanar arrangement of the cells if the nanofibers were highly oriented [169].

Degradation of smooth muscle cell cultured PGA, PLGA, and P(LLA-CL) nanofibers were investigated [170]. It has been found that cell culture accelerated the nanofiber scaffold degradation, especially for PGA. P(LLA-CL) nanofibers facilitate long-term (1–3 months) cell growth because of the slowest degradation rate.

2.1.6 Skins

As the largest body tissue, skin functions to prevent outside intrusion and regulate water retention and heat loss. Nanofiber mats from many different types of polymers have been evaluated for their cytocompatibility to fibroblast and/or keratinocytes. The polymers studied include natural polymers, such as silk fibroin [171], collagen [172], gelatin [173], hyaluronic acid (HA) [174], chitosan [140], and synthetic polymers such as PLGA [133], PCL [175], PLLA [176], and polyethersulfone (PES) [177]. In most cases, the electrospun nanofiber mats exhibited good capability of supporting cell attachment and proliferation. Cells seeded on the nanofiber structure tended to maintain phenotypic shape and were guided to grow according to the nanofiber orientation.

Collagen-PCL core-sheath nanofibers were prepared by both one-step coaxial electrospinning and two-step spinning and solution-coating process [175, 178]. The presence of collagen surface shell on PCL nanofibers was found to support cell proliferation and encourage cell migration inside the scaffold. The collagen coating formed directly from the coaxial-electrospinning had higher cell proliferation efficiency than that from the solution-coating method.

Nanofibers of polymer blends were also studied, and showed good cell attachment and proliferation. It was indicated that a small fraction of low molecular weight polyethylene glycol (PEG) in PLLA nanofibers increased the hydrophilicity of the nanofibers. As a result, biological reactivity of fibroblast cells was improved [176]. The fiber wettability could also be improved by co-electrospinning of two different nanofibers (PLGA and chitosan/PVA) into the same fibrous matrix [140, 179]. The presence of hydrophilic nanofibers (chitosan/PVA) increased the absorption of nutrient fluid during cell culture and thus promoted fibroblast attachment, proliferation, migration, and infiltration in the fiber matrix. In addition, grafting nanofiber surface with a hydrophilic polymer also improved the wettability of nanofibers, an example being the treatment of PLLA, PLGA, and PGA nanofibers with oxygen plasma followed by a chemical reaction with acrylic acid [180, 181]. PCL can also be used to improve mechanical properties of collagen nanofibers, but PCL concentration is critical and should be kept lower than 30% to maintain the high-level epidermal formation and cell viability [182].

In addition to improve the surface wettability, bioactivity of nanofiber matrices was also enhanced by functionalizing nanofiber surface with a cell adhesive peptide [183]. The cell attachment, spreading, and proliferation were greatly enhanced in the modified nanofiber mat compared with the un-modified control.

The effect of nanofiber alignment on cell adhesion and proliferation was studied [184]. When a rabbit conjunctive fibroblast was seeded on an aligned collagen nanofiber mat, lower cell adhesion but higher cell proliferation was observed on the aligned nanofibers, compared with that on the randomly orientated nanofiber mat. It has also been observed that the infiltration of epidermal skin cells was enhanced when aligned PLLA nanofibers were used [185].

Cell culture on a 3D nanofiber mat containing large pores was also studied. A 3D silk fibroin nanofiber matrix was prepared by directly depositing the newly electrospun silk fibroin nanofibers in a methanol solution [186]. This 3D nanofiber matrix supported the cell proliferation inside the fiber matrix quite well. A 3D nanofiber mat was also prepared by adding a chemical blowing agent (BA) into nanofibers via electrospinning, and a post-electrospinning heat-treatment, leading to large pores within the nanofiber mat [187]. These large pores could provide the cells with good access to nutrients inside the nanofiber mat. In addition, in a 3D cell culture system using aligned fiber array, the maximum fiber diameter for cell adhesion and the maximum inter-fiber distance bridging cell aggregation were demonstrated to be 10 and 200 μm, respectively [188]. The relationship between inter-fiber distance and skin tissue regeneration has been investigated on electrospun gelatin scaffolds [189]. Cell infiltration was found only in the upper regions of the scaffold if the inter-fiber distance was smaller than 5.5 μm, and the distances between 5 and 10 μm have yielded most favorable skin substitute.

2.1.7 Neural Tissues

Electrospun nanofibers represent effective guidance substrates for nervous tissue repair. In vitro cell culture study of neural stem cells

(NSCs) on a PLLA nanofiber scaffold revealed that the nanofiber scaffold not only supported the NSC differentiation and neurite outgrowth, but also promoted NSC adhesion [190]. Further work using aligned PLLA nanofiber or microfiber scaffolds showed that the directions of NSC elongation and its neurite outgrowth were parallel to the direction of fiber alignment [191]. The aligned fibers in the scaffold can provide topographical guidance and enhance cell proliferation and neurite outgrowth [192], which has the potential for bridging long peripheral gaps.

So far, the effects of fiber diameter on cell culturing are still unclear. PLLA nanofibers had higher neural stem cell differentiation than microfibers, but with similar cell orientation [191]. Among three kinds of PES nanofibers with the diameters of 283, 749, and 1452 nm, cell proliferation and spreading increased with the decrease in fiber diameter [193]. In another research on three kinds of PLLA electrospun fibers with different diameters, the neurite length was the shortest on the finest fibers [194].

In comparison with pure PCL nanofibers, collagen/PCL nanofibers improved Schwann cell migration, neurite orientation, and process formation of Schwann cells, fibroblasts, and olfactory ensheathing cells. Fiber surface tension has been adjusted by treatment under a base condition [195] or surface coating [196] to achieve increased cell number and spreading. Also, in vitro response of Schwann cells on different types of polymer nanofibers, such as poly(3-hydroxybutyrate) (PHB), PHBV, PCL, PLLA, and chitosan, and their solution-cast films, was examined [197].

Electrical stimulation has been incorporated during cell growth to improve cell function by preparing conductive nanofibrous scaffolds [198, 199]. PPy was grown on random and aligned PLGA nanofibers, and the electrically stimulated nanofibers resulted 40–50% longer neuritis and 40–90% more neurite formation compared to the one without any stimulation.

The feasibility of in vivo nerve regeneration using PLGA nanofibers was also investigated [200, 201]. After implantation of PLGA nanofiber guidance to the right sciatic nerve of rats, no inflammatory response was observed, and 5 out of 11 rats showed successful nerve regeneration 1 month after implantation. The in vitro test also confirmed that nerve stem cells adhered to, and differentiated on, the PLGA nanofiber mats [202].

2.1.8 Other Tissue Scaffolds

In addition to the above-mentioned scaffolds, nanofiber scaffolds have been used to support other stem cells [203, 204], and tissues, such as heart [205], ligament [206], and urinary tract [207].

2.1.9 Wound Healing

Wound healing is a native process of regenerating dermal and epidermal tissues. When an individual is wounded, a set of complex biochemical actions take place in a closely orchestrated cascade to repair the damage. These events can be classified into inflammatory, proliferative, and remodeling phases and epithelialization. Normally, body cannot heal a deep dermal injury. In full thickness burns or deep ulcers, there is no source of cells remaining for regeneration, except from the wound edges. As a result, complete re-epithelialization takes a long time and is complicated with scarring of the base [208].

Dressings for wound healing function to protect the wound, exude extra body fluids from the wound area, decontaminate the exogenous microorganism, improve the appearance, and sometimes accelerate the healing process. For these functions, a wound-dressing material should provide a physical barrier to a wound, but be permeable to moisture and oxygen. For a full thickness dermal injury, the adhesion and integration of an "artificial dermal layer" consisting of a 3D tissue scaffold with well-cultured dermal fibroblasts considerably assists in the re-epithelialization.

Electrospun nanofiber mat is a good wound dressing candidate because of its unique properties: the highly porous mat structure and well-interconnected pores are particularly important for exuding fluid from the wound; the small pores and very high-specific surface area not only inhibit the exogenous microorganism invasions, but also assist the control of fluid drainage; in addition, the electrospinning process provides a simple way to add drugs into the nanofibers for any possible medical treatment and antibacterial purposes.

A study on using electrospun PU mat as wound-dressing material revealed that the mat effectively exuded fluid from the wound, without fluid accumulation under the mat, and no wound desiccation occurred either [209]. The mat also showed a controlled water loss from evaporation, excellent oxygen permeability, and high fluid drainage ability, besides inhibiting the invasion of exogenous microorganism. Histological test indicated that the rate of epithelialization was increased and the dermis became well organized when the wounds were covered with the electrospun nanofiber mat. Silk nanofiber mats were electrospun with epidermal growth factor to promote wound-healing process [210], and it has been found that the functionalized mat increased the wound closure by 90%. An anti-cancer, anti-oxidant, and anti-flammatory substance, curcumin, was loaded into PCL [211] and CA [212] nanofibers for wound dressing. These dressings were superior in supporting human dermal fibroblast attachment and proliferation and exhibited high rate of wound closure.

An open-wound-healing test using an electrospun collagen nanofiber mat showed that the early stage healing using collagen nanofiber mat was faster than that of using normal cotton gauze [213]. In the first week, the wound surface in the cotton group was covered by fibrous tissue debris, below which dense infiltration of polymorphonuclear leukocytes and the proliferation of fibroblasts were formed. By comparison, the surface tissue debris in the collagen nanofiber group disappeared, and prominent proliferation of young capillaries and fibroblasts was found. Later stage healing processes were similar for both groups. PLGA/collagen nanofiber mats also showed an effective wound healing behavior with an active response to human fibroblasts in the early stage [214].

For dermal wound healing, electrospun nanofiber mats have been used as scaffolds to mimic native extracellular matrix. To achieve this objective, extensive studies have been conducted to develop biocompatible electrospun nanofibrous scaffolds for culturing dermal fibroblasts [215, 216].

In vivo wound healing of diabetic ulcers was investigated using electrospun block copolymer (PCL-PEG) and PCL. When the nanofibers were chemically modified with a recombinant human epidermal growth factor (rhEGF), the expression of keratinocyte-specific genes and EGF-receptor were enhanced [217]. Systematic in vivo wound healing evaluations have been carried on PVA, PCL, PAN, PVDF-HFP, PAN/PEU blend, and wool protein coated PVA and PCL nanofibers [218, 219]. The results showed that wound healing performance was mainly influenced by the porosity, air permeability, and surface wettability of the nanofiber mats. A mat with good hydrophilicity and high porosity facilitated the healing at the early stage. However, the fiber diameter and antibacterial activity had a limited effect on the healing efficiency.

Post-surgery tissue adhesion is a widely recognized problem for abdominal surgeries. It not only renders future operations more difficult but also causes other problems such as small bowel obstruction, female infertility, and chronic debilitating pain [220, 221]. An electrospun nanofiber mat containing antibiotic agents has been used as a barrier to prevent the post-surgery abdominal adhesions. It was found that the nanofiber mat eliminated post-surgery abdominal adhesion significantly, thus improving the healing process [222].

To decontaminate the bacteria invasion, biocides, such as silver [223–226] and iodine complex [227], have been added to the electrospun nanofibers. It was reported that PVP-iodine complex gradually released active iodine. Because of the broad-spectrum microbicidal activity of the iodine, electrospun PVP-iodine nanofibers had external antibacterial, antimycotic, and antiviral applications. Silver (Ag) ion was reported to have a biocidal effect on as many as 16 types of bacteria including Escherichia coli (Ec) and Staphylococcus aureus (Sa) [228]. Ag ions were incorporated into electrospun nanofibers via adding $AgNO_3$ into the polymer solution for electrospinning. To maintain a long-term antibacterial activity and control the release of Ag ions, the Ag was embedded in the form of elementary state by a post-electrospinning treatment. Ag nanoparticles can also be directly incorporated into electrospun nanofibers via the electrospinning process. An Ag/PVA nanofiber mat exhibited excellent antimicrobial ability and good stability in moisture environment, as well as a quick and continuous release with good effectiveness [223].

Besides adding antibacterial additives, antimicrobial nanofibers can also be prepared by directly using antimicrobial polymers. For instance, polyurethanes containing quaternary ammonium groups were electrospun into nanofiber nonwovens, and the nanofibers showed very strong antimicrobial activities against Sa and Escherichia coli [229].

Some other electrospun materials, such as Ag containing hydrogel [225], hyperbranched polyglycerol [230], and PES [177], have also been investigated as wound dressing.

2.1.10 Drug Delivery and Release Control

Controlled release is an efficient process of delivering drugs in medical therapy. It can balance the delivery kinetics, immunize the toxicity and side effects, and improve patient convenience [231]. In a controlled release system, the active substance is loaded into a carrier or device first, and then released at a predictable rate in vivo when administered by an injected or non-injected route.

As a potential drug delivery carrier, electrospun nanofibers have exhibited many advantages. The drug loading is very easy to implement via electrospinning process, and the high applied voltage used in the electrospinning process had little influence on the drug activity. The high specific surface area and short diffusion passage length give the nanofiber drug system higher overall release rate than the bulk material (e.g., film). The release profile can be finely controlled by modulation of nanofiber morphology, porosity, and composition.

Nanofibers for drug release systems mainly come from biodegradable polymers and hydrophilic polymers. Model drugs that have been studied include water-soluble [229, 232], poor water-soluble [233, 234], and water-insoluble drugs [235]. The release of macromolecules, such as DNA [236–238] and bioactive proteins [239, 240], from nanofibers was also investigated. Many factors influence the release behavior, such as the type of polymers used, wettability of drugs and polymers, solubility, drug–polymer comparability, additives, and the existence of enzyme in the buffer solution. In addition, the rate of releasing a water-soluble drug could be slowed down when nanofiber matrix was crosslinked.

In most cases, water-soluble drugs, including DNA and proteins, exhibited an early-stage burst [232, 241]. For some applications, preventing post-surgery-induced adhesion for instance, and such an early burst release will be an ideal profile because most infections occur within the first few hours after surgery. A recent study also found that when a poorly water-soluble drug was loaded into PVP nanofibers [242], and 84.9% of the drug can be released in the first 20 s when the drug-to-PVP ratio was kept as 1:4, which can be used for fast drug delivery systems. However, for a long-lasting release process, it would be essential to maintain the release at an even and stable pace, and any early burst release should be avoided. For a water-insoluble drug, the drug release from hydrophobic nanofibers into buffer solution is difficult. However, when an enzyme capable of degrading nanofibers exists in the buffer solution, the drug can be released at a constant rate because of the degradation of nanofibers [235, 243]. For example, when rifampin was encapsulated in PLA nanofibers, no drug release was detected from the nanofibers. However, when the buffer solution contained proteinase K, the drug release took place nearly in zero-order kinetics, and no early burst release happened. Similarly, initial burst release did not occur for poor water-soluble drugs, but the release from a non-biodegradable nanofiber could follow different kinetics [233]. In another example, blending a hydrophilic but water-insoluble polymer (PEG-g-CHN) with PLGA could assist in the release of a poor water-soluble drug Ibuprofen [244]. However, when a water-soluble polymer was used, the poorly soluble drug was released accompanied with dissolving of the nanofibers, leading to a low burst release [245, 246]. By loading both hydrophilic and hydrophobic drugs into same nanofiber mat [247], the release of the hydrophilic drug could accelerate the release of the hydrophobic drug.

The early burst release can be reduced when the drug is encapsulated within the nanofiber matrix. When an amphiphilic block copolymer, PEG-b-PLA was added into Mefoxin/PLGA nanofibers, the cumulative amount of the released drug at earlier time points was reduced and the drug release was prolonged [241]. The reason for the reduced burst release was attributed to the encapsulation of some drug molecules within the hydrophilic block of the PEG-b-PLA. Amphiphilic block copolymer also assisted the dispersion and encapsulation of water-soluble drug into nanofibers when the polymer solution used an oleophilic solvent, such as chloroform, during electrospinning [248]. In this case, a water-in-oil emulation can be electrospun into uniform nanofibers, and drug molecules are trapped by hydrophilic chains. The swelling of the hydrophilic chains during releasing assists the diffusion of drug from nanofibers to the buffer.

Coating nanofibers with a shell could be an effective way to control the release profile. When a thin layer of hydrophobic polymer, such as poly (p-xylylene) (PPX), was coated on PVA nanofibers loaded with bovine serum albumin (BSA)/luciferase, the early burst release of the enzyme was prevented [239]. Fluorination treatment [249] on PVA nanofibers introduced functional C-F groups and made the fiber surface hydrophobic, which dramatically decreased the initial drug burst and prolonged the total release time.

The polymer shell can also be directly applied via a coaxial co-electrospinning process, and the nanofibers produced are normally named "core-sheath" bicomponent nanofibers. In this case, even a pure drug can be entrapped into nanofiber as the core, and the release profile was less dependent on the solubility of drug released [250–253]. A research has compared the release behavior of two drug-loaded PLLA nanofibers prepared using blend and coaxial electrospinning [254]. It was found that the blend fibers still showed an early burst release, while the threads made of core-sheath fibers provided a stable release of growth factor and other therapeutic drugs.

In addition, the early burst release can also be lowered via encapsulating drugs into nanomaterial, followed by incorporating the drug-loaded nanomaterials into nanofibers. For example, halloysite nanotubes loaded with tetracycline hydrochloride were incorporated into PLGA nanofibers and showed greatly reduced initial burst release [255].

2.2 Environmental Protection

Human activities in the last 150 years have had serious negative impacts on our environment, and this is likely to continue for a very long time. Effective solutions are urgently needed to protect our environment. Due to their high specific surface area, electrospun nanofibers are expected to be used to collect pollutants via physical blocking or chemical adsorption.

2.2.1 Filtration

Electrospun nanofibers for filtration application have a long history. A company in the USA (Donaldson) has produced electrospun nanofiber-based filter products for industry, consumer, and defense applications for more than 20 years, and its Ultra-web® nanofiber filter has been developed for nonwoven and filtration industry for a wide range of applications. Recently another company (AMSOIL) has also developed a nanofiber-based fuel filter for automobile applications. DuPont has electrospun fabric products for heating, ventilation, and air conditioning (HVAC), automotive and liquid filtration, bedding protection, and apparel applications.

In military, nanofiber filters are used in uniform garments and isolating bags to decontaminate aerosol dusts, bacteria, and even virus. Respirator is another example that requires efficient filtration ability. Similar function is also needed for some fabrics used in the medical area.

For a fiber-based filter, the removal of particles is determined by different mechanisms. Large particles are blocked on the filter surface due to the sieve effect. Particles that are smaller than the surface-pores could still be collected by the fibers, through a mechanism such as interception, impaction, or static electrical attraction. Very fine particles could also be captured by the Brownian motion effect. The filtration efficiency is normally influenced by the filter physical structure (fiber fineness, matrix structure, thickness, and pore size), fiber surface electronic properties, and its surface chemical characteristic (e.g., surface free energy). The particle-collecting capability is also related to the size range of particles being collected. Besides the filtration efficiency, other properties such as pressure drop and flux resistance are also important factors to be evaluated for a filter media.

Electrospun nanofiber mat provides dramatic increases in filtration efficiency at relatively small decreases in permeability. In comparison with conventional filter fibers at the same pressure drop, nanofibers with a diameter finer than half a micron have a much higher capability to collect the fine particles, because the slip flow around the nanofibers increases the diffusion, interception, and inertial impaction efficiencies [256].

Both experimental measurements and theoretical calculations revealed that electrospun nanofiber mats were extremely efficient at trapping airborne particles (0.5–200 μm) [257]. A very thin layer of electrospun nanofibers sprayed onto a porous substrate was sufficient to eliminate the particle penetration. The air-flow resistance and aerosol filtration properties correlate with the add-on weight of the electrospun fiber coating. Also, electrospun layers present minimal impedance to moisture vapor diffusion, which is very important for protective clothing in decontamination applications. A comparison study between a nylon-6 electrospun mat (thickness 100 μm, pore size 0.24 μm) and a commercial high efficiency particulate air (HEPA) filter (thickness 500 μm, pore size 1.7 μm) using 300 nm test particles indicated that the thin nanofiber mat had a slightly higher filtration efficiency (99.993%) than the HEPA filter (99.97%) [258].

Besides solid particles, tiny liquid droplets within a liquid–liquid immiscible system could also be removed by a nanofiber mat (liquid–liquid coalescence filtration). PS nanofibers (diameter about 600 nm) were electrospun from a recycled expanded-polystyrene (EPS), and mixed with micro glass fibers to form a filter media for removal of water droplets from a water-in-oil emulsion [259]. The addition of small amount of PS nanofibers was reported to significantly improve the capture efficiency (from 68% to 88%), but the pressure drop of the filters was increased considerably. In another trial, electrospun nylon nanofibers were also blended with glass fibers (diameter 5 μm) for the coalescence filtration, and addition of an optimal amount of nanofibers (1.6 wt%) to the coalescence filter improved the capture efficiency, but did not cause excessive pressure drop [260].

Nanofibers were used as a supporting scaffold in ultrafiltration (UF) for oil/water emulsion separation. The reported UF mat has a three-layered composite structure consisting of a nonporous hydrophilic top layer, a crosslinked PVA electrospun nanofibrous mid-layer, and a conventional nonwoven microfibrous substrate [261, 262]. The electrospun nanofibrous substrate provided a well-interconnected porous network with a large specific surface area. The UF filter has a high flux rate and excellent organic solute rejection capability.

Long-term filtration performance can be extended by using chemical [263] or physical [264] crosslinking treatments to stabilize the porous structure of nanofibrous mat. It was recently reported that filters made of multiple thin nanofiber layers had a much better filtration performance than single thick-layer nanofiber mat in terms of the pressure drop during filtration [265].

Other works related to the filtration applications include effects of operating parameters in electrospinning on fiber morphology and pore structure for filter media [266], nanofiber mat for separation of micro particles from liquid [267], pre-filter nanofiber mat for particulate removal [268], and water filtration [269].

2.2.2 Metal Ion Adsorption and Recovery

The elevated level of metal ions has become a serious pollution in water resources, and gives a long-term risk to human health and natural environment. These positively charged ions can be removed by many different methods, and among them adsorption through chemical and physical affinity is the most effective and economic one. In this case, electrospun nanofibers have great potential in collecting metal ions from a solution because of their high specific surface area, high porosity, and controllable surface functionality.

Polymers with functional groups that have affinity to metal ions were directly electrospun into nanofibers for metal ion adsorption. Electrospun wool keratin/silk fibroin blend nanofibers were initially used to chelate absorb Cu(II) ions from water [270]. Wool protein has a large number of hydrophilic amino acids with high affinity to metal ions, and the blend nanofibers showed a Cu(II) adsorption capability of 2.8 μg/mg with high recycling efficiency. Chitosan [271] and chitosan/PEO-blend [272] nanofibers have also exhibited high adsorption of Cu(II), Pb(II), and Cr(VI).

Two approaches have been used to improve the adsorption of metal ions on electrospun nanofibers: introducing functional materials to fiber surface using surface chemistry or coating techniques and increasing surface area to improve adsorption capability. For example, amidino diethylenediamine chelating groups was incorporated into PAN nanofiber using heterogeneous reaction with diethylenetriamine [273], and the modified fibers had a pH-dependent adsorption capacities of 150.6, 155.5, 116.5, and 60.6 mg/g for Cu(II), Ag(I), Fe(II), and Pb(II), respectively. The Cu(II) and Pb(II) adsorption capacities of amidoxime-modified PAN nanofibers were 52.70 and 263.45 mg/g [274]. Boehmite nanoparticles were incorporated into PCL and nylon-6 nanofibers for Cd(II) collection [275]. In recent research [276], PVA/SiO_2 composite nanofibers were prepared and functionalized by mercapto groups. The specific surface area of the nanofibers was higher than 290 m^2/g, and the nanofibers exhibited the highest Cu(II) adsorption capacity of 489.12 mg/g.

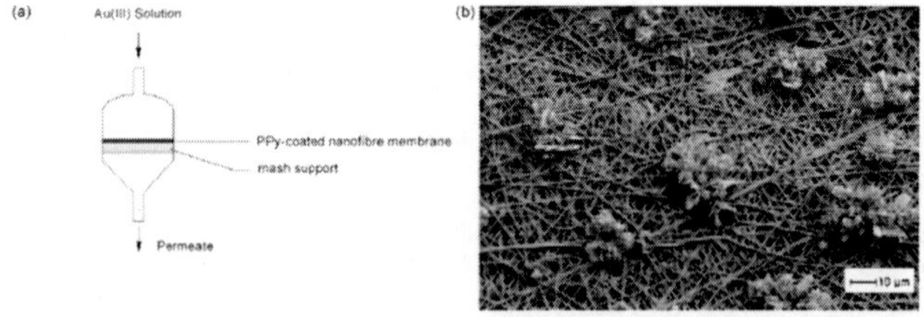

Figure 2.3 (a) Apparatus for mat-based Au recovering, (b) a SEM image of the PPy-coated nanofibers with aqueous $[Au(III)Cl_4]^-$ solution permeated [277].

Apart from the adsorption mechanism, directly converting metal ions into metal element on nanofibers was used for recovering precious metals. Wang et al. [277] reported a simple and interesting method to grow polypyrrole on nanofiber surface using vapor-phase polymerization. A thin layer of PPy coating on the nanofibers had very little effect on the permeability of the nanofiber membrane, but showed a high efficiency in recovering gold ions. The gold ions were converted into micron-sized gold particles (Fig. 2.3).

2.3 Catalyst and Enzyme Carriers

A carrier for catalyst in chemistry and biology is used to preserve high catalysis activity, increase the stability, and simplify the reaction process. An inert porous material with a large surface area and high permeability to reactants could be a promising candidate for efficient catalyst carriers. Using an electrospun nanofiber mat as a catalyst carrier, the extremely large surface could provide a huge number of active sites, thus enhancing the catalytic capability. The well-interconnected small pores in the nanofiber mat warrant effective interactions between the reactant and catalyst, which is valuable for continuous-flow chemical reactions or biological processes. The catalyst can also be grafted onto the electrospun nanofiber surface via coating or surface modification.

2.3.1 Catalysts

Palladium (Pd)-loaded poly(acrylonitrile-acrylic acid) (PAN-AA) nanofibers were prepared by electrospinning a $PdCl_2$-containing PAN-AA solution into nanofibers and subsequently reducing the $PdCl_2$ embedded into Pd nanoparticles. The static catalytic activity of the Pd-loaded nanofibers for selective hydrogenation of dehydrolinalol was measured to be about 4.5 times higher than that of Pd/Al_2O_3 catalyst [278]. In a similar work, Pd-loaded PAN-AA nanofibers were confirmed to have high activity and good recycling property for hydrogenation of a-olefin at room temperature. The yield of the hexene to hexane catalyzed by the Pd/PAN-AA nanofibers was 4.7 times higher than that of $Pd/\gamma\text{-}Al_2O_3$ [279].

When a molecular catalyst was incorporated into nanofibers, the catalyst could leak out of the nanofibers during the catalysis reaction [280]. Surface coating such as catalyst-loaded nanofiber with a thin layer of polymer considerably retained the catalyst in the nanofibers and the catalyst efficiency was improved at the same time. The catalysis performance was influenced by the type of shell polymer and coating thickness.

A study on using polyacrylamidoxime (PANOx)-containing PAN nanofibers for hydrolysis of p-nitro-phenyl acetate (PNPA) indicated that diffusion resistance among the nanofibers might limit the accessibility of oxime catalytic sites in the fiber and thus influence the catalytic activity [281].

TiO_2 is a widely used catalytic material in the commercial market. Electrospun PVA-platinum (Pt)/TiO_2 nanofibers were prepared for photocatalytic degradation of solid-state PVA [282]. Compared with TiO_2 nanoparticles, electrospun TiO_2 nanofibers showed higher photocatalytic activities in terms of photocurrent generation by a factor of three because of the mesoporosity and nanoparticle alignment, which caused efficient charge separation through interparticle charge transfer along the nanofiber framework [283]. Fiber diameter was reported to play an important role in determining the photocatalytic activity of TiO_2 nanofibers, and the diameter of 200 nm was found to be optimal in photocatalytic performance [284].

Doping has been used to further increase the photocatalytic activity of TiO_2 nanofibers. Materials such as Pt [285], gold [286], SnO_2 [287], vanadium [288], and Fe_2O_3 [289] have been used to dope TiO_2 nanofibers for heterostructure using different approaches, and enhanced photocatalytic behavior was achieved. SnO_2 with different morphologies and densities was formed on TiO_2 nanorods (Fig. 2.4a–d). These SnO_2/TiO_2 heterostructures were demonstrated to have high photocatalytic degradation efficiency, and the photocatalytic performance was dependent on the SnO_2 morphology (Fig. 2.4e).

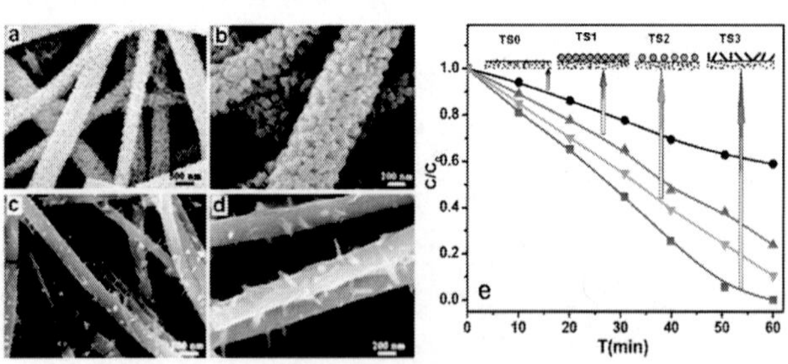

Figure 2.4 (a–d) SEM images of different SnO_2 morphologies on TiO_2 nanofibers and (e) the SnO_2-morphology-dependent photocatalytic degradation behavior of Rhodamin B [287].

Apart from TiO_2 nanofibers, other inorganic nanofibers (such as ZnO [290] and $LaCoO_3$ [291]) have also been investigated as catalysts. For example, ZnO-based hierarchical nanofibers have been prepared by growing radically oriented ZnO nanowires on electrospun PLLA nanofibers to make flexible catalyst for continuous-flow photocatalytic water purification [292].

2.3.2 Enzymes

Chemical reactions using enzymes as catalysts have high selectivity and require mild reaction conditions. For easy separation from the reaction solution, enzymes are normally immobilized with a carrier. The immobilization efficiency mainly depends on the porous structure and enzyme–matrix interaction. Nano-structured materials are recently used as enzyme carriers because of their large specific surface area and the high loading capability [293]. To immobilize enzyme on electrospun nanofibers, many approaches have been used, including grafting enzyme on fiber surface, physical adsorption, and incorporating enzyme into nanofiber via electrospinning followed by crosslinking reaction.

To graft enzymes on nanofiber surface, the polymer used should possess reactive groups for chemical bonding [294–299]. In some studies, polymer blends containing at least one reactive polymer were used [300–302]. The immobilized enzymes normally showed a slightly reduced activity in aqueous environment compared with the un-immobilized native counterpart, but the activity in non-aqueous solution was much higher. For example, α-chymotrypsin was used as a model enzyme to bond chemically on the surface of electrospun PS nanofibers. The enzyme was measured to cover over 27.4% monolayer of the nanofiber surface, and the apparent hydrolytic activity of the enzyme-loaded was 65% of the native enzyme, while the activity in non-aqueous solution was over 3 orders of magnitude higher than that of its native enzyme under the same condition. In another study using PAN nanofibers to immobilize lipase, the tensile strength of the nanofiber mat was improved after lipase immobilization, and the immobilized lipase retained >90% of its initial reactivity after being stored in a buffer at 30°C for 20 days, whereas the free lipase lost 80% of its initial reactivity. Also the immobilized lipase still retained 70% of its specific activity after 10 repeated reaction cycles [297]. In addition, the immobilized enzyme also showed improved pH and thermal stabilities [298]. Ethylenediamine was used to modify PAN nanofiber mat to introduce active and hydrophilic groups, followed by a chitosan coating for improvement of biocompatibility [303].

Enzymes were incorporated into nanofibers via electrospinning, and subsequent crosslinking the enzymes incorporated effectively prevented their leaching. In the presence of PEO or PVA, casein and lipase were electrospun into ultra-thin fibers. After crosslinking with 4,4'-methylenebis(phenyl diisocyanate) (MDI), the fibers became insoluble, and the lipase en- capsulated exhibited six times higher hydrolysis activity towards olive oil than that of the films cast from the same solution [304]. The crosslinked enzymes in nanofibers showed very high activity and stability. For example, the immobilized α-chymotrypsin in a shaken buffer solution maintained the same activity for more than 2 weeks [305].

In addition to chemical bonding, the enzymes were also applied onto nanofibers simply via physical adsorption [306, 307]. Polyacrylonitriles-2-methacryloyloxyethyl phosphoryl choline (PANCMPC) nanofiber was reported to have high biocompatibility with enzymes because of the formation of phospholipid micro-environment on the nanofiber surface. Lipase on the nanofibers showed a high immobilization rate, strong specific activity, and good activity retention.

2.4 Sensors

Sensors have been widely used to detect chemicals for environment protection, industrial process control, medical diagnosis, safety, security, and defense applications. A good sensor should have a small dimension, low fabrication cost, and multiple functions, besides the high sensitivity, selectivity, and reliability [308]. High sensitivity and fast response require the sensor device having a large specific surface area and highly porous structure. The characteristics possessed by electrospun nanofibers match well with these requirements. Therefore a nanofibrous structure should be a promising physical structure to form a highly sensitive and fast response sensor.

Several approaches have been used to impart nanofibers with a sensing capability, such as using a polymeric sensing material to electrospin nanofibers, incorporating sensing molecules into nanofibers, or applying sensing material on nanofiber surface via coating/grafting technique.

Electrospun poly(acrylic acid) (PAA) nanofibers have been grafted with pyrene methanol (PM) as the sensing material to detect metal ions Fe^{3+} and Hg^{2+}, and an explosive 2,4-dinitrotuloene (DNT) in water [309, 310]. Due to the quenching effect of these chemicals to the pyrene moieties, the fluorescent intensity of nanofibers had a linear response to the concentration of quenchers, and the nanofibers showed high sensitivities. Similarly, a PM-grafted PMMA nanofibers showed an order of magnitude higher sensitivity to target analyte DNT than its cast film counterpart [311]. Explosive 2,4,6-trinitrotoluene (TNT) vapor can be accurately detected using electrospun PAN nanofibers that were coated with a thin layer of conjugated polymer, poly(triphenylamine-alt-biphenylene vinylene) (TPA-PBPV) [312]. Highly porous electrospun SnO_2 nanofibers showed extremely high sensitivity to ethanol (< 1ppb) after etching with an oxygen plasma [313].

Fluorescence optical sensors were also prepared by a layer-by-layer electrostatic assembly technique to apply a conjugated polymer onto nanofiber surface for detection of methyl viologen and cytochrome c in aqueous solution [314], and porphyrin-doped silica nanofibers were used to trace TNT vapor [315]. All those nanofiber sensors showed high sensitivity and rapid response. Besides fluorescent properties, conjugated polymer-embedded electrospun nanofibers were also reported to be able to sense volatile organic compounds (VOCs) based on optical absorption properties [316].

A gas sensor using a specific absorption interaction between ammonia and PAA nanofibers was reported. The weight difference induced by the gas absorption was measured by a quartz crystal microbalance (QCM). This sensor was capable of detecting ppb level NH_3 in air, and the sensitivity was four times higher than that of the PAA cast film [317]. The absorption of gas also leads to changes in Fourier transform infrared spectroscopy (FTIR) absorption. PAN nanofibers containing metal oxide nanoparticles, such as iron oxide and zinc oxide, have been used to detect carbon dioxide [318]. The addition of metal oxide nanoparticles enhanced the gas adsorption and thus improved the sensitivity. In an electrospun carbon nanofiber gas sensor [319], carbon black was blended into the fibers to increase the conductivity, and a porous structure was introduced to the fiber surface by etching under a basic condition. The sensory ability for NO_2 and CO gases was reported to be improved dramatically, and the enhanced sensitivity was attributed to higher surface area and improved electrical conductivity due to the formation of carbon black network. It has been found that electrospun SnO_2 fibers with a smaller grain size showed much better sensitivity to CO and NO_2 gases. This was probably because the fibers containing smaller grains had higher resistance, which was the result of larger number of grain–grain interconnections [320].

Electrical conductivity is an important property for sensor devices. Conducting nanofibers can be produced from semi-conducting oxides, conducting polymers, and non-conductive polymers. Pure oxide nanofibers are normally produced by electrospinning a solution containing oxide sol-gel and polymer, followed by calcining treatment to remove the polymer. The detection of gas molecules using oxide nanofibers is based on the conductivity changes due to the doping effect of analyst gases to the oxide. A few oxide nanofibers have been assessed for detecting different gases, such as MoO_3 [321] nanofibers for ammonia, WO_3 nanofibers for ammonia [322] and NO_2 [323], TiO_2 nanofibers for NO_2 and H_2 [324]. These sensors exhibited improved sensitivity, faster response, and lower detection limit than that of sol-gel based films. In addition, Co-doped ZnO nanofibers were used as photoelectric oxygen sensor with fast response [325].

Conducting polymer is another interesting sensor material. Electrospun PANi/PS nanofibers containing glucose oxidase have been demonstrated to have a high sensitivity to glucose [326]. PANi/PVP nanofibers also exhibited sensing ability to NO_2 [327]. Organic/inorganic semiconductor Schottky nanodiode was fabricated by PANi nanofibers and inorganic n-doped semiconductor. The device has a rapid response and supersensitive to ammonia [328]. Poly(3,4-ethylenedioxythiophene)-poly(styrene sulfonate) (PEDOT:PSS) has also been blended into PVP nanofibers for chemical vapor detection [329].

Besides the conducting and semi-conducting materials, insulating polymers were also used to fabricate electrical sensors. In this case, ions or conductive nano-fillers were added to improve the conductivity. When nanofibers of PEO were doped with $LiClO_4$, the mat showed low conductivity and was sensitive to moisture, and the nanofiber mat was reported to have much higher sensitivity than their film-type counterpart [326]. Carbon nanotubes/PVDF composite nanofibers showed an increased straining sensing ability (as measured by voltage across the sensor), 35 times higher than that of the film counterpart [330]. In addition, electrospun nanofibers incorporated with carbon black showed sensitivity to VOCs [331]. When the carbon black concentration was near the percolation threshold, the composite fibers changed their resistance in volatile organic compounds. Using different polymer matrices, the sensor can be used to detect toluene, trichloroethylene, methanol, and dichloropentane vapors.

Humidity sensor is a very important device for environment tests. KCl doped ZnO [332] and TiO_2 [333] nanofibers have shown higher humidity sensitivity with faster response and recovery time compared to pure ZnO and TiO_2 nanofibers. $BaTiO_3$ nanofibers also exhibited excellent humidity sensing behavior because its complex impedance varied around three orders of magnitude in the whole humidity range [334].

In addition, nanofiber-based electrochemical sensors were also reported. For example, PVdF/poly(aminophenylboronic acid) (PAPBA) nanofibers reacted with glucose, and the change in amperometric properties was detected easily [335].

2.5 Energy Harvest and Storage

Energy is essential for our modern civilization. The rapidly growing global energy demand has not only sped up the consumption of non-renewable fossil fuels, but also threatens regional stability. In addition, to reduce the energy consumption using highly efficient technology, converting other energy into electrical power can considerably assist in alleviating the energy crisis. In this direction, nanotechnology is providing new solutions to solve the problems. It has been found that nanofibrous materials can have significantly higher energy conversion and storage efficiency than their bulk counterparts.

2.5.1 Solar Cells

Solar cells utilize unlimited solar energy for power generation and have been considered as a major solution to current energy crisis. So far, single crystal and polycrystalline silicon based solar cells are dominating the commercial solar cell market. Dye-sensitized solar cells (DSSCs) and organic solar cells are still under development.

Conventional DSSCs have a dye-anchored mesoporous TiO_2 nanoparticle thin layer sandwiched between two conducting glass plates in the presence of an electrolyte. Since electrospun nanofibers are one-dimensional material with better electrical conductivity and higher specific surface area than nanoparticles, and large pore size in nanofibrous mat allows increased penetration of viscous polymer gel electrolyte, nanofibers have shown great application potential in DSSCs.

By using a thin layer of TiO_2 nanofibers as working electrode in DSSC, the photocurrent generation using polymer gel electrolytes was over 90% of liquid electrolyte [336, 337]. DSSC devices [338, 339] using polymer electrolytes based on electrospun PVDF-HFP nanofibers could achieve higher conversion efficiency and better long-term stability than conventional liquid electrolyte ones.

The largest problem of using inorganic nanofibrous layers in DSSC is the poor adhesion with conductive substrate. The nanofiber layers are easily separated after calcination due to different thermal expansion coefficients. Many approaches have been tested to address this issue. A TiO_2 nanoparticle layer was coated on conductive glass before nanofiber deposition to increase the adhesion between the glass and the TiO_2 layer [340]. Hot pressing pre-treatment was applied on ZnO/PVA composite nanofiber mat before calcination [341]. A self-relaxation layer [342] was also spontaneously formed during electrospinning to release the interfacial tensile stress generated in calcination.

To increase the short-circuit current, the electrospun TiO_2 electrode was treated with $TiCl_4$ aqueous solution to form an additional rutile TiO_2 layer on the fiber surface. Such a rutile TiO_2 layer increased the fraction volume of active TiO_2 and inter-fiber connection, resulting in an increased photocurrent [336, 343, 344]. Another method to improve charge generation and transport was through materials doping. Quantum dots (QDs) were decorated onto TiO_2 nanofibers to yield multiple carrier generation due to the quantum confinement effect [345, 346]. The electron diffusion coefficient and mobility of the niobium-doped anatase TiO_2 nanofibers were an order of magnitude higher than those of the un-doped fibers [347].

Another advantage of electrospun nanofibers is that they can be used to form a transparent mat for counter electrode. This effort has become more and more urgent because of the rapidly growing price of tin and indium, which are typically used for making conductive glass. Copper nanofiber thin layers were prepared using electrospinning for transparent electrode in DSSC [348]. The resulted fibers had ultrahigh aspect ratios of up to 100,000 and fused crossing points with ultra-low junction resistance. The fibrous mat exhibited great flexibility and stretch ability as well. Electrospun carbon nanofibers were also explored as counter electrode material, and an energy conversion efficiency of 5.5% was achieved in the resultant solar cell [349]. The carbon nanofiber based DSSC had lower fill factor and overall performance, because of the higher total series resistance ($15.5\ \Omega\cdot cm^2$) than that ($4.8\ \Omega\cdot cm^2$) of Pt-cased traditional conductive substrate.

2.5.2 Fuel Cells

Fuel cells are electrochemical devices capable of converting hydrogen or hydrogen-rich fuels into electrical current by a metal catalyst. There are many kinds of fuel cells, such as proton exchange mat (PEM) fuel cells, direct methanol fuel cells, alkaline fuel cells, and solid oxide fuel cells [350]. PEM fuel cells are the most important one among them because of high power density and low operating temperature.

Pt nanoparticle catalyst is a main component in fuel cells. The price of Pt has driven up the cell cost and limited the commercialization. Electrospun materials have been prepared as alternative catalyst with high catalytic efficiency, good durability, and affordable cost. Binary PtRh and PtRu nanowires were synthesized by electrospinning, and they had better catalytic performance than commercial nanoparticle catalyst because of the one-dimensional features [351]. Pt nanowires also showed higher catalytic activities in a polymer electrolyte membrane fuel cell [352].

Instead of direct use as catalyst, catalyst-supporting material is another important application area for electrospun nanofibers. Pt clusters were electrodeposited on a carbon nanofiber mat for methanol oxidation, and the catalytic peak current of the composite catalyst reached 420 mA/mg compared with 185 mA/mg of a commercial Pt catalyst [353]. Pt nanoparticles were immobilized on polyimide-based nanofibers using a hydrolysis process [354], and Pt nanoparticles were also loaded on the carbon nanotube containing polyamic acid nanofibers to achieve high catalytic current with long-term stability [355].

Proton exchange mat is the essential element of PEM fuel cells and normally made of a Nafion film for proton conduction. Because pure Nafion is not suitable for electrospinning due to its low viscosity in solution, it is normally mixed with other polymers to make blend nanofibers. Blend Nafion/PEO nanofibers were embedded in an inert polymer matrix to make a proton conducting mat [356], and a high proton conductivity of 0.06–0.08 S/cm at 15°C in water and low water swelling of 12–23 wt% at 25°C were achieved. Besides blend electrospinning, Nafion surface coating on polymeric nanofibers [357–359] is another efficient way to obtain better fuel cell performance and reduce Nafion consumption.

2.5.3 Mechanical Energy Harvesters

Piezoelectric material can directly convert mechanical energies into electrical power and has shown great potential in powering low energy consumption devices. Piezoelectric power generators, also called nanogenerators in some literature, are becoming an important source for renewable energy [360, 361]. These nanogenerators are normally made of aligned inorganic nanowires and their preparation required preciously controlled conditions [362, 363].

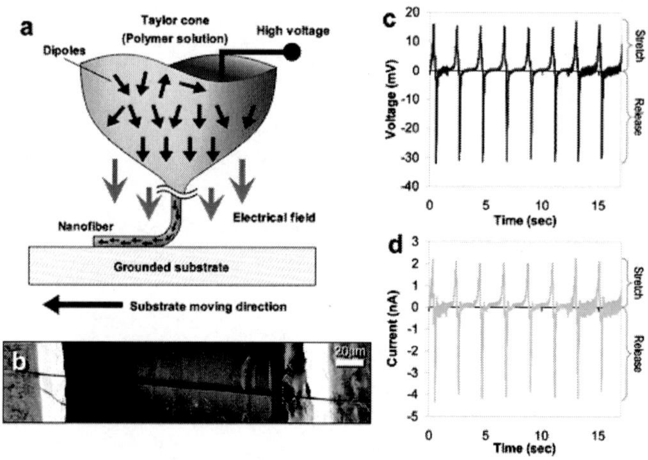

Figure 2.5 (a) Scheme of near-field electrospinning process; (b) a SEM image of a single PVdF nanofiber power generator; (c) and (d) voltage and current outputs from the generator under applied strain at 2 Hz [364].

Electrospinning can process different piezoelectric materials into nanofibrous structure and electrospun piezoelectric nanofibers have been demonstrated to have energy scavenging capability. PVDF nanofiber was directly deposited across two metal electrodes without extra poling treatment to make single-fiber nanogenerator using a near-field electrospinning technique (Fig. 2.5a,b) [364]. Voltage (5–30 mV) and current (0.5–3 nA) outputs were generated when the fiber underwent repeated extensive deformation at 2 Hz (Fig. 2.5c,d). Rigid lead-zirconate-titanate (PZT) material has been made into electrospun nanofibers with highly improved flexibility for fabricating flexible power generator after a soft polymer packaging [365]. The output voltage and power under repeated stress application were 1.63 V and 0.03 μW, respectively. Aligned PVdF nanofibers were collected using a normal electrospinning process, and the following poling treatment was used to improve piezo-responsiveness [366]. A hybrid nano system combining this fiber-based power generator and a flexible enzymatic biofuel cell was demonstrated as a self-powered UV light sensor.

Because of the existence of strong electric field and fiber stretching during electrospinning, randomly oriented PVDF electrospun nanofiber membranes were recently reported to prepare piezoelectric power generators [367]. By simply sandwiching a small piece of nanofiber membrane (2 cm^2) between two Al foil electrodes and pressing the device (Fig. 2.6), voltage as high as several volts was generated and the output was impact frequency dependent. And voltage outputs were converted to DC signals using a full-wave rectifier bridge and stored in capacitors to continuously light a LED light (Fig. 2.7). At almost the same time, Mandal et al. also confirmed the preferential orientation of CF_2 dipoles in piezoelectric poly(vinylidene fluoride-trifluoroethylene) (P(VDF-TrFE)) nanofiber membranes and demonstrated their potential application as pressure sensor [368]. Electrospinning provides a simple but effective method to prepare piezoelectric nanofibrous membranes with potential applications in self-powered microelectronics and wearable electronics.

2.5.4 Lithium Ion Batteries

Lithium ion batteries are compact and rechargeable electrical energy storage devices. They have very high energy density. Like most other chemical batteries, porous structure is an essential requirement for the battery. A sponge-like electrode has high discharge current

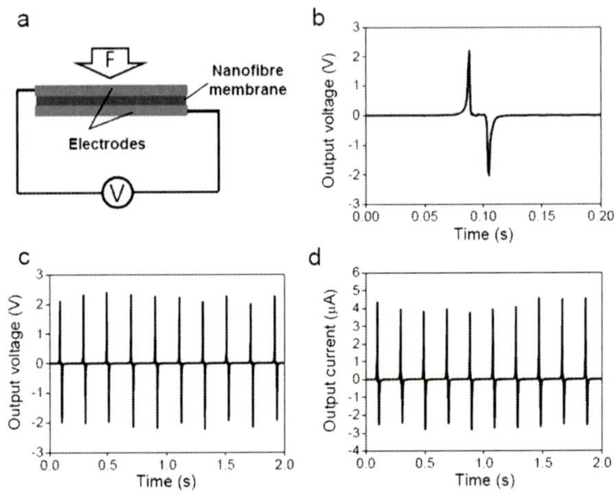

Figure 2.6 (a) Illustration of power-generating device; (b) typical voltage outpus of a nanofiber membrane device subjected to an impact action (initial impact speed, 34 mm s^{-1}); (c) voltage and (d) current outputs of the same nanofiber device under 5 Hz repeated compressive impacts (nanofiber membrane thickness = 140 μm; working area = 2 cm^2) [367].

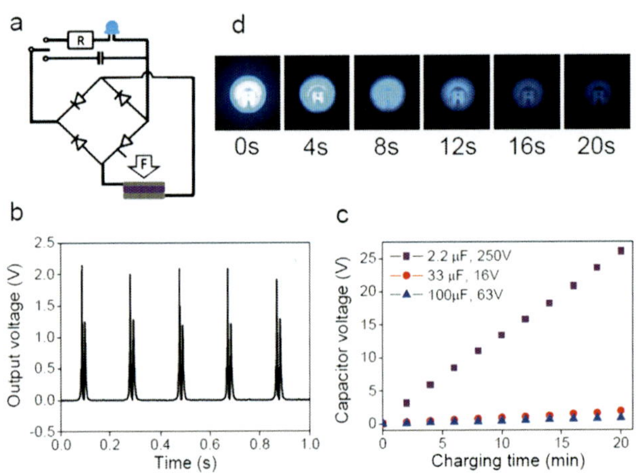

Figure 2.7 (a) Circuit for charging a capacitor and lighting a LED device; (b) DC voltage output rectified from a nanofiber membrane device; (c) voltage-charging-time relationship for a nanofiber membrane device loaded with different capacitors; (d) still frames captured from a video showing continuous lighting of a blue LED powered by a nanofiber power generator [367].

and capacity, and a porous separator between the electrodes can effectively stop the short circuit, but allows the exchange of ions freely. Solid electrolytes used in lithium ion battery are typically composed of a gel or porous host to retain the liquid electrolyte inside [369]. To have high ion conductivity, the host material, also called separator, should have high permeability to ions. A porous mat with well-interconnected pores, suitable mechanical strength, and high electrochemical stability could be a potential candidate.

Perfluoro polymers have been widely studied in polymer electrolyte because of their chemical and mechanical stability. An electrospun PVDF polymer electrolyte showed high uptake to electrolyte solution (320–350%) and high ion conductivity (1.7×10^{-3} S/cm at 0°C). The fibrous electrolyte also had high electrochemical stability of more than 5 V. The prototype cell (mesocarbon microbeas (MCMB)/PVDF-based electrolyte/$LiCoO_2$) exhibited a very stable charge–discharge behavior [370]. When a thin layer of polyethylene (PE) was plasma polymerized onto PVDF nanofiber surface, a role of shutter by melting of the PE layer grafted was rendered to the nanofiber mat and improved the battery safety [371]. It was also found that the formation of interconnected web structure via heat treatment improved both the mechanical properties and dimensional stability of nanofiber mats [371, 372].

Nanofiber mats from other types of polymers, such as PAN [373], were also studied as a lithium battery separator. The electrospun PAN mat showed high ion conductivity and electrochemical stability. The prototype cell based on the electrospun PAN electrolyte separator with 1 M $LiPF_6$-EC/DMC exhibited an initial discharge capacity of 145 mAh/g, and 94.1% of the initial discharge capacity after 150 cycles at a charge/discharge rate of 0.5 C/0.5 C. To improve ionic conductivity, charge–discharge capability and stability, SiO_2 nanoparticles [374], polydiphenylamine (PDPA) [375], PMMA [376], and PAN [377] were blended with PVDF-HFP nanofibers for making composite polymer electrolyte.

Besides being used as a separator, some electrospun nanofiber mats have been used as battery electrodes. For example, a carbonated electrospun nanofiber mat was used as an anode in a lithium ion battery, and the batteries showed a large reversible capacity of 450 mAh/g [378]. One recent research has focused on improving battery performance by incorporating different inorganic nanomaterials into carbon nanofibers as electrodes. An anode material made of

Fe_3O_4/C composite nanofibers had much better electrochemical performance with a high reversible capability of 1007 mAh/g at the 80th cycle and excellent rate capability [379]. A Sn/C composite was encapsulated into hollow carbon nanofibers as an anodematerial for lithium batteries [380] with a high reversible capacity of 737 mAh/g after 200 cycles at 0.5 C (480 mAh/g at 5°C).

Also, $LiNi_{1/3}Mn_{1/3}Co_{1/3}O_2$ [381], vanadium oxide [382], and $LiFePO_4$/carbon composite nanofibers [383] have been investigated as cathode material for lithium ion batteries.

2.5.5 Supercapacitors

As an electrochemical device with a high power density and super high charging–discharging rate, supercapacitors (also known as double-layer capacitor) have been demonstrated with great potential in different emerging applications, such as power back-up for laptop or mobile phone and power source for hybrid electric vehicle. For a supercapacitor with high capacitive behavior, electrodes made of porous carbon materials are extremely important, and electrospinning technique has been used to prepare carbon nanofibrous mats with high specific surface area and controllable pore size for this purpose. Activated carbon nanofibers prepared from a PAN/DMF solution have shown a maximum specific capacitance of 173 F/g [384]. The PBI-based carbon nanofibers had specific surface areas ranging from 500 to 1220 m^2/g, and the fabricated double-layer capacitor exhibited specific capacitance between 35 and 202 F/g [385].

The capacitance of electrospun nanofiber-based supercapacitor can be enhanced by using composite carbon electrodes. $ZnCl_2$ [386], carbon nanotubes [387], and nickel [388] have been blended into procurer solutions for electrospinning and then carbonization. The $ZnCl_2$/C composite nanofibers exhibited a capacitance of 140 F/g with a specific surface area of 55 m^2/g when the carbon nanofibers were doped with 5 wt% $ZnCl_2$. The specific capacitance of an electrical double-layer capacitor with electrodes made of carbon nanotubes embedded carbon nanofiber reached as high as 310 F/g.

In a recent publication [389], bicomponent electrospinning was applied to prepare side-by-side nanofibers with PAN on one side and a thermoplastic polymer, PVP, on the other side. The resultant carbon fibers after pyrolysis showed improved inter carbon fiber connections and crystallization. The capacitance of the electrochemical cell (Fig.

2.8a) made of these inter-bonded carbon nanofibers was much higher than that of the carbon nanofibers from polymer blends (Fig. 2.8b,c).

Apart from carbon, other inorganic nanofiber electrodes have also been examined for supercapacitor applications. For example, RuO_2 has been deposited on a Pt nanofiber mat to function as hybrid electrode, and the supercapacitor exhibited a specific capacitance of 409.4 F/g with a capacity loss of only 21.4% from 10 to 1000 mV/s [390]. Electrospun V_2O_5 nanofiber-based supercapacitor had the highest specific capacitance of 190 F/g in an aqueous electrolyte and 250 F/g in an organic electrolyte when the nanofibers were annealed at 400°C [391].

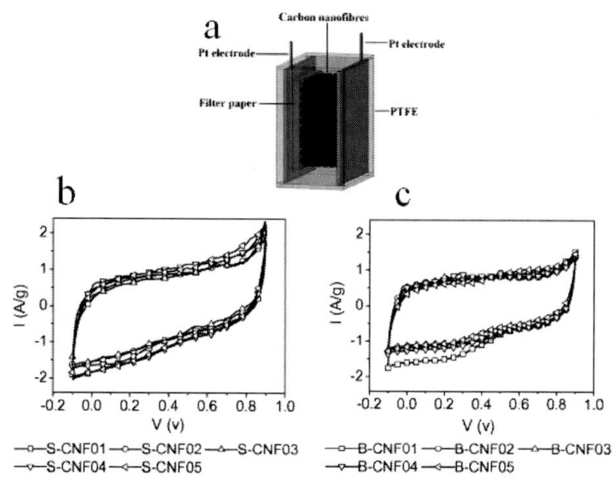

Figure 2.8 (a) Illustration of electrochemical cell and the cyclic voltammetry curves of side-by-side (b) and blend (c) carbon nanofibers [389].

2.5.6 Hydrogen Storage

Hydrogen has been widely known as an ideal alternative energy to solve energy crisis and global warming [392]. Because high-pressure and cryogenic storage systems can't satisfy the criteria for on-board storage, the key issue for the current hydrogen energy research is hydrogen storage [393].

It has been reported that carbonaceous materials, such as active carbon, carbon nanotubes, and graphite, are able to store hydrogen because of their large specific surface area, high pore volume, and light weight. The hydrogen-storage capabilities of electrospun carbon fibers carbonated from different starting polymers (PAN [394], PVDF [395], PANi [396], and polycarbosilane (PCS) [397]) have also been assessed. Carbon nanofiber can be a better hydrogen storage material than other carbon materials because it has an optimized pore structure with controlled pore size [398].

To increase graphitization during carbonation, Fe (III) acetylacetonate was used as a catalyst to prepare graphite nanofibers, and the resulting nanofiber (surface areas of 60–253 m^2/g) had the H-storage capacity of 0.14–1.01 wt%. To increase the specific surface area of carbon nanofibers, different inorganic materials (normally metal [399] and metal oxides [400, 401]) were mixed into polymer solutions for electrospinning. The resultant carbon nanofiber could have a specific surface area of 2900 m^2/g with an H-storage capacity of 3 wt%. Even though hydrogen adsorption ability increases with the increase in specific surface area, pore volume is also very important. It has been concluded that the most effective pore width is in the range 0.6–0.7 nm [402], which is slightly larger than the hydrogen molecule (0.4059 nm).

Exciting results have been obtained more recently. PANi fibers were prepared by electrospinning and showed a reversible hydrogen storage capacity of 3–10 wt% at different temperatures [396]. Highly porous carbide-derived carbon fibers with a specific surface area of 3116 m^2/g were prepared after pyrolysis and chlorination of electrospun PCS fibers [397]. The fibers have shown a very high hydrogen storage capability of 3.86 wt% at a low pressure (17 bar).

2.6 Other Applications

Early studies on electrospun nanofibers also included reinforcement of polymers. As electrospun nanofiber mats have a large specific surface area and an irregular pore structure, mechanical interlocking among the nanofibers should occur. When a thin electrospun nylon-4,6 nanofiber mat was added to epoxy, the composite showed transparency to visible light, and both the stiffness and strength were increased considerably compared with the pure epoxy film [403]. Recently, polysulfone (PSF) [404] and polyetherketone cardo

(PEK-C) [405] nanofibers were used to improve the toughness of carbon fiber/epoxy composite. When carbon nanofibers were dispersed into the PEK-C nanofiber phase, a synergistic effect to enhance the toughness and other mechanical properties was also observed [406].

Electrospun PBI nanofibers have been used as fillers to reinforce epoxy and rubber [407]. An epoxy containing 15 wt% electrospun PBI nanofibers was found to have higher fracture toughness and modulus than the one containing 17 wt% PBI whiskers. Also the Young's modulus and tear strength of styrene-butadiene rubber (SBR) containing the PBI nanofibers were higher than those of pure SBR. In addition, electrospun nylon PA 6 nanofibers were used to improve the mechanical properties of BISGMA/TEGDMA dental restorative composite resins [408].

Electrospun nanofibers showed excellent capability to absorb sound. A leading nanofiber technology company, Elmarco, recently patented an electrospun nanofiber material that had unique sound absorption characteristics, with only about one-third of the weight of conventional sound absorption materials. It was able to absorb sounds across a wide range of frequencies, especially those below 1000 Hz.

Electrospun nanofibers have also shown application potential in the field-effect transistor (FET). EFT behavior has been observed in camphor sulfonic-acid-doped electrospun PANi/PEO nanofibers [409]. Saturation channel currents were found at low-source-drain voltage with a hole mobility in the depletion regime of 1.4×10^{-4} cm^2/Vs. Electrospun nanofiber mat has also been demonstrated to allow for ultrafast identification of latent fingerprints [410].

Chapter 3

Developments in Large-Scale Nanofiber Electrospinning Systems

Needle electrospinning has been used extensively in research labs to produce a small quantity of nanofiber samples or as a model in theoretical analysis to understand the basic electrospinning process. Electrospinning systems with large-scale nanofiber production abilities have been explored, and some inspiring results have been achieved. This chapter summarizes the recent research progress in large-scale electrospinning technologies.

3.1 Downward Multi-Jet Electrospinning

A straightforward way to increase the electrospinning throughput is to use multi-jet spinnerets as shown in Fig. 3.1a. The fiber productivity can be simply increased by increasing the jet number [411, 412]. However, multi-jet electrospinning has shown strong repulsion among the jets, and this may lead to reduced fiber production and poor fiber quality, which is the main obstacle for its practical application. To reduce the jet repulsion, jets have to be set at an appropriate distance apart, and a large space is required to accommodate the needles for the mass nanofiber production.

To stabilize and optimize the electrospinning process, an extra-cylindrical electrode has been used as an auxiliary electrode to cover the multi-jet spinneret [413]. As shown in Fig. 3.1b, the presence of

Needleless Electrospinning of Nanofibers: Technology and Applications
Tong Lin and Xungai Wang

ISBN 978-981-4316-35-4 (Hardcover), 978-981-4316-84-2 (eBook)
www.panstanford.com

the external electrode dramatically reduced the fiber deposition area, thus improving the fiber production rate. Nevertheless, coarser fibers were observed, because the auxiliary electrode shortened the chaotic motion of multi-jets in electrospinning.

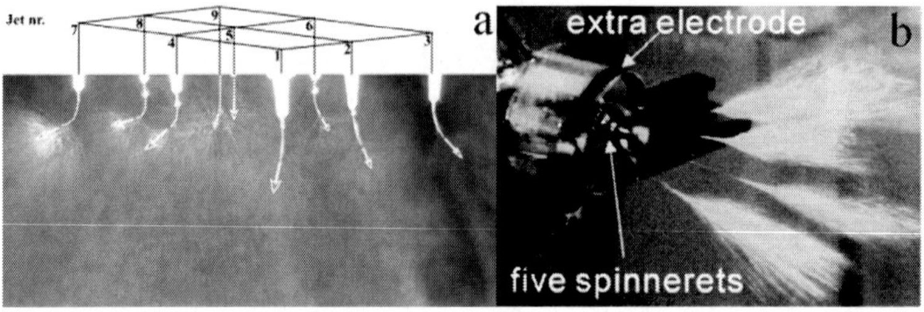

Figure 3.1 (a) A multi-jet electrospinning setup [414], (b) multi-jet electrospinning with a cylindrical auxiliary electrode [413].

In electrospinning, the ejected solution jet carries a large amount of charge, which drive the jet stretching and fiber deposition on the collector. According to the electrospinning mechanism, it can be deduced that interferences in multi-jet electrospinning are unable to be eliminated completely, which will be a barrier to the industrialization of multi-jet electrospinning. Furthermore, the successful operation of multi-jet electrospinning requires a regular cleaning system to avoid the blockage of the needle nozzles. Setting the cleaning device for each needle makes it almost impossible to use multi-jet electrospinning for the mass production of nanofibers.

In addition to the multi-jet electrospinning, porous tubes have also been used as electrospinning spinneret to improve the fiber productivity. This system is herein still classified under multi-jet electrospinning because multiple fibers are simultaneously produced from polymer solution channels inside the tube. A porous polyethylene tube with a vertical axis was used to electrospin nanofibers (Fig. 3.2a) [415]. The production rate was reported to be 250 times greater than that of single needle electrospinning. However, the SEM results of obtained nanofibers showed large variations in fiber diameter.

In another example of tube electrospinning (horizontal tube), the polymer solution was pushed through the tube wall with many holes at 1–2 kPa pressure (Fig. 3.2b) [416]. This setup can only produce 0.3–0.5 g/h of nanofiber due to the small number of holes (fiber generators) that can be drilled per unit area. Although it was mentioned that the production rates can be easily scaled up by increasing the tube length and the number of holes, the space between holes can't be reduced much because of the electric field repulsion between the jets. The strong jet interference in this setup can even result in a nanofiber belt instead of fiber web [416] .

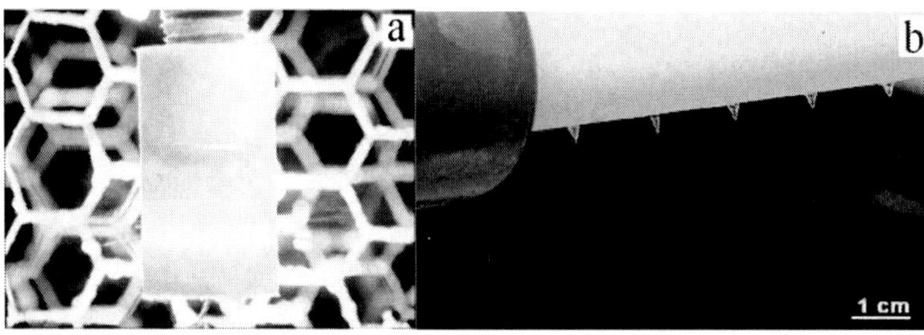

Figure 3.2 (a) Electrospinning using a vertical tubular foam spinneret [415], (b) horizontal tube electrospinning [416].

Wang et al. reported a conical wire coil electrospinning spinneret, which can work at up to 70 kV without causing corona discharges [417]. As shown in Fig. 3.3a–c, the spinning solution was held in the wire cone without using any solution channels. Due to the large surface tension and visco-elasticity, the polymer solution can be retained inside the wire cone. When a high electric voltage was applied to the wire coil, solution was stretched out from the wire surface and the gap between wires to form solution jets. Without using defined solution channels, the fiber ejected independently, eliminating the limitation of jet number that was formed in multi-jet electrospinning and tube electrospinning. In comparison with conventional needle electrospinning, wire coil electrospinning can improve the electrospinning productivity noticeably and produce nanofibers with smaller fiber diameter (Fig. 3.3d). The results also indicated that fibers prepared by coil electrospinning had a wider diameter distribution than those produced by the needle electrospinning.

An edge-plate electrospinning setup for improving the fiber production rate was reported by Thoppey et al. [418]. A plate ($\theta = 40°$) for retaining solution was used as a spinneret to electrospin polyethylene oxide (PEO) nanofibers (Fig. 3.4a). In this method an electrically insulated reservoir connected with one or more plastic pipettes supplied solution to the charged plate, and each pipette supplied a solution stream at the jet initiation site. It was found that the production rate was increased by over five times even using a single spinning site (one pipette) without getting coarser fibers and wider diameter distribution (Fig. 3.4b,c). The surface tension of polymer solution plays a vital role, and the plate angle must be set appropriately according to the solution properties, otherwise solution dripping may occur.

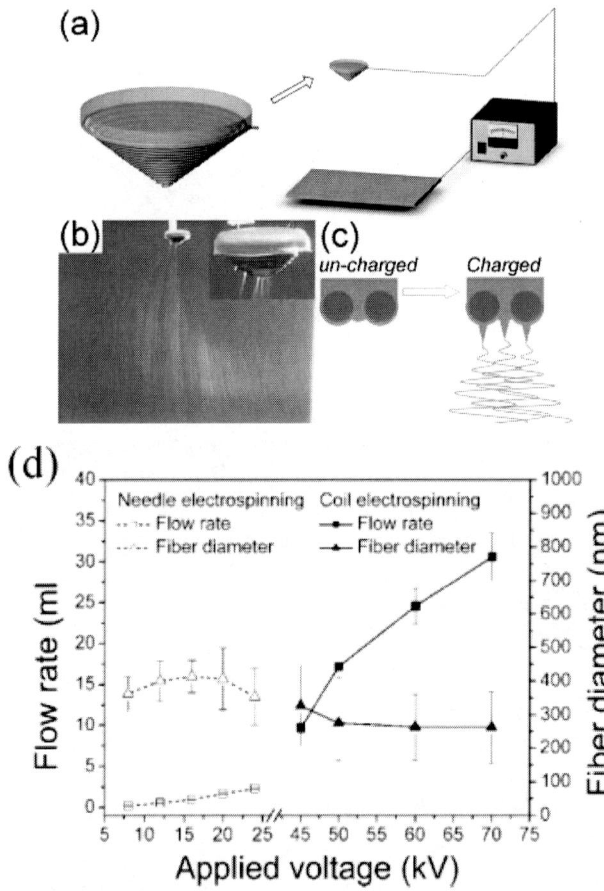

Figure 3.3 (a) Schematic illustration of conical wire coil electrospinning setup, (b) photograph of the electrospinning process, (c) illustration of jet formation, (d) comparison between needle electrospinning and coil electrospinning (PVA concentration = 9 wt%; collecting distance = 15 cm) [417] .

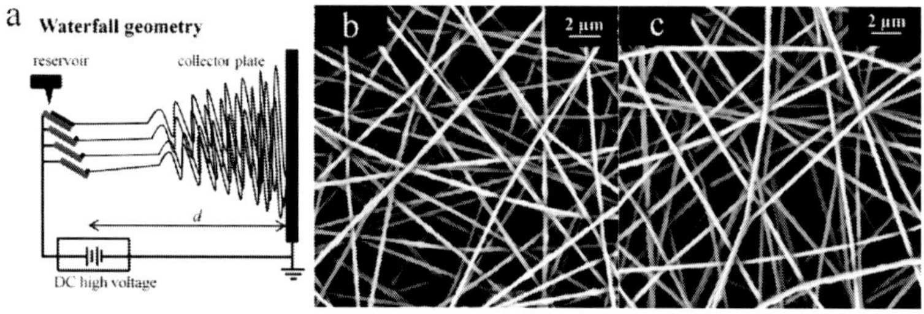

Figure 3.4 (a) Schematic illustration of plate edge electrospinning, (b) and (c) SEM images of nanofibers electrospun from (b) conventional needle electrospinning (collecting distance = 15 cm, applied voltage = 11 kV) and (c) edge-plate geometry (collecting distance = 35 cm, applied voltage = 28 kV) [418].

In another electrospinning design, a solution reservoir was used to provide spinning solution to a metal roller electrospinning spinneret (Fig. 3.5) [419]. The polymer solution droplets were splashed onto the surface of a metal roller by a solution distributor, which had a hole at the bottom. When the voltage was applied, solution droplets adhering on the surface of metal roller spinneret were ejected and stretched under the electric force to form nanofibers. This setup was proposed to have the ability to perform electrospinning with improved fiber production rate.

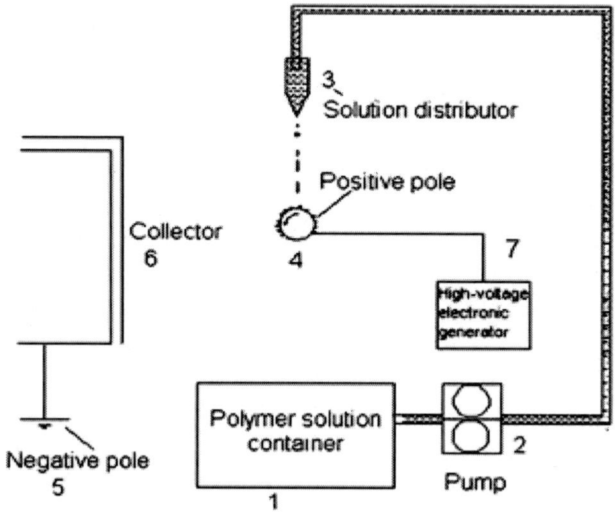

Figure 3.5 Schematic illustration of splashing electrospinning setup [419].

Recently, a rotary cone was used as an electrospinning spinneret to perform electrospinning, which used a glass pipe to supply the PVP solution to the cone to ensure enough solution for continuous electrospinning (Fig. 3.6a) [420]. The electrospinning throughput of this setup was reported to be 1000 times larger than that of conventional needle electrospinning. The morphologies of nanofibers prepared by the cone electrospinning were nearly the same as those produced by conventional needle electrospinning (Fig. 3.6b,c).

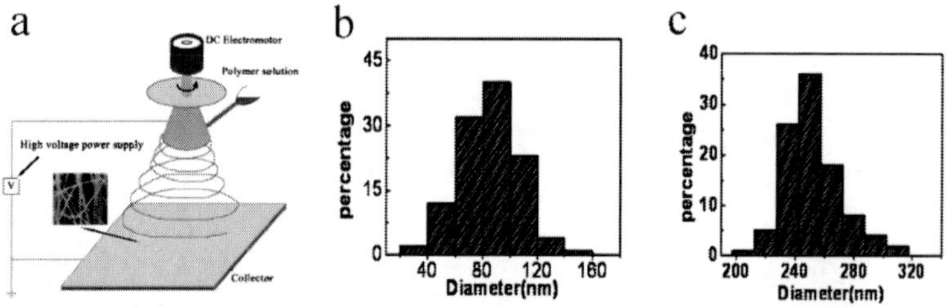

Figure 3.6 (a) Schematic illustration of the rotary cone electrospinning setup (inset: SEM image of collected PVP nanofibers, rotational speed of cone = 100 rpm, applied voltage = 30 kV, collecting distance = 20 cm, solution throughput = 10 g min^{-1}), (b) fiber diameter distribution of needle electrospinning, (c) fiber diameter distribution of rotary cone electrospinning [420].

3.2 Upward Needleless Electrospinning

3.2.1 Electrospinning Techniques

A good electrospinning method suitable for manufacturing nanofibers should have minimal dependence on the fluidic channel numbers to improve the fiber productivity. It should be universal for processing polymer solutions of different properties. Some upward needleless electrospinning setups have good potential.

Yarin and Zussman [421] reported a two-layer-fluid electrospinning setup (Fig. 3.7a) that could dispose of the problems related to multi-jet electrospinning. In this setup, the lower fluid layer was a ferromagnetic suspension and the upper layer was the polymer solution to be spun. During electrospinning, when a normal magnetic field was applied to the system, steady vertical spikes were formed perturbing the interlayer interface. As a result of applying a high voltage to the fluid at the same time, thousands of jets ejected upward (Fig. 3.7b). This upward electrospinning system required a complicated setup and the resultant nanofibers had large fiber diameter and wide diameter distribution.

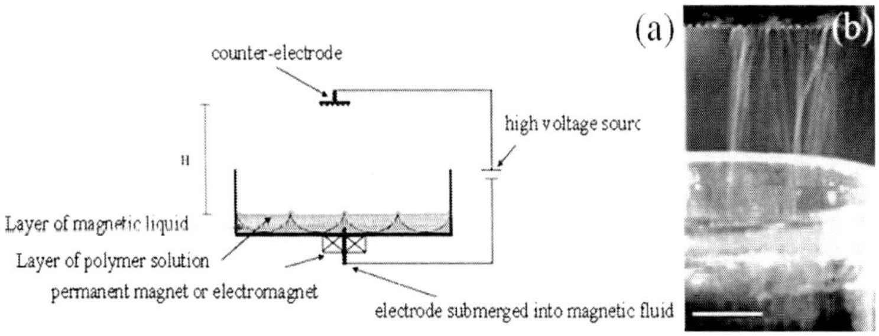

Figure 3.7 (a) A two-layer-fluid electrospinning setup, (b) multiple jets ejected toward the counter-electrode.

In another upward electrospinning work, bubbles or humps were generated on the free surface of a polymer solution to initiate the electrospinning process [422]. Unlike the previous design, a high pressure gas was used by inserting a gas tube to the bottom of the solution reservoir. A flat aluminum plate was used as collector above the solution. When a high voltage was applied to the solution, Taylor cones were easily formed from the humps. The fiber production rate was reported to depend on the gas pressure, the solution property, and the applied voltage. However, fibers prepared by this method contained large beads.

Jirsak et al. [423] invented a needleless electrospinning setup by using a rotating roller as fiber generator. When the roller was partially immersed into a polymer solution while rotating slowly, the polymer solution was loaded onto the upper roller surface. Upon applying a high voltage to the electrospinning system, an enormous number of solution jets can be generated from the roller surface upward (Fig. 3.8). This setup has been commercialized by Elmarco Co with the brand name "Nanospider^{TM}."

Lukas et al. [424] developed a one-dimensional electrohydrodynamic theory to describe the electrospinning of conductive liquids from an open flat surface based on the phenomenon that nanofibers can be electrospun from linear clefts even without being aided by a magnetic fluid underneath (Fig. 3.9).

Figure 3.8 Roller electrospinning process (left), and commercialized Nanospider^{TM} (right) [423].

This work added a general approach toward studying dynamics of surface waves. During electrospinning from a free liquid surface, due to the electric force the amplitude of a characteristic wavelength boundlessly grew faster than the others. The fastest growing stationary wave marked the onset of electrospinning from a free liquid surface with polymer jets originating from the wave crests. The proposed theory predicated the critical field strength and corresponding critical inter-jet distance, and the inter-jet distance for field strengths above the critical value. The theory also predicted relaxation time necessary for spontaneous jetting after a high voltage was applied, and explained the fundamental of upward needleless electrospinning.

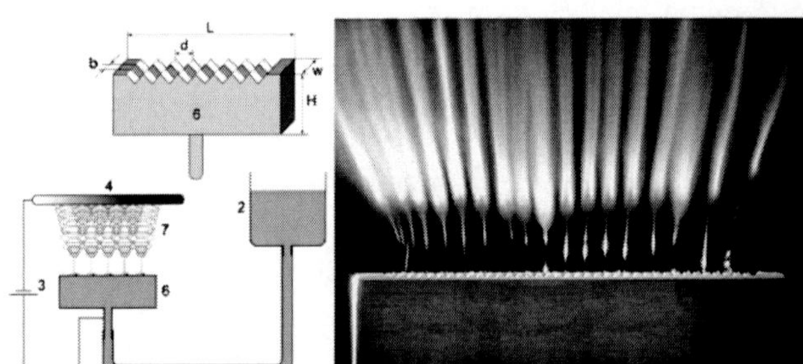

Figure 3.9 Schematic illustration of a linear cleft electrospinning setup (left) and electrospinning process (right) [424].

Needleless electrospinning process with a rotating spinneret can be summarized as that, the rotation of spinneret loads a thin layer of polymer solution onto the spinneret surface. The rotation and gravity create conical spikes on the surface of this solution layer. When a high voltage is applied to the spinneret, these spikes tend to concentrate charges and amplify the perturbance, and the fluid around the spikes is drawn to these spikes under high electric force. Taylor cones are thus formed. Fine solution jets are then ejected from the tips of these Taylor cones, when the electric force is large enough.

Using a similar design, Niu et al. systematically compared the needleless electrospinning using different rotary fiber-generators (disc, cylinder, or ball) [425] (Fig. 3.10). When PVA solution was charged with a high electric voltage via a copper wire inside the solution vessel, numerous jets/filaments were generated from the spinnerets, which deposited on the collector (e.g., rotating drum). With the rotation of the spinneret, PVA solution was loaded onto the spinneret surface constantly, leading to continuous generation of polymer jets/filaments. They also used finite element method to analyze electric field and examine the influence of spinneret shape on the electric field intensity profile. They found that the spinneret with a highly concentrated electric field overlapping the fiber generating area is the key to efficient needleless electrospinning of uniform nanofibers.

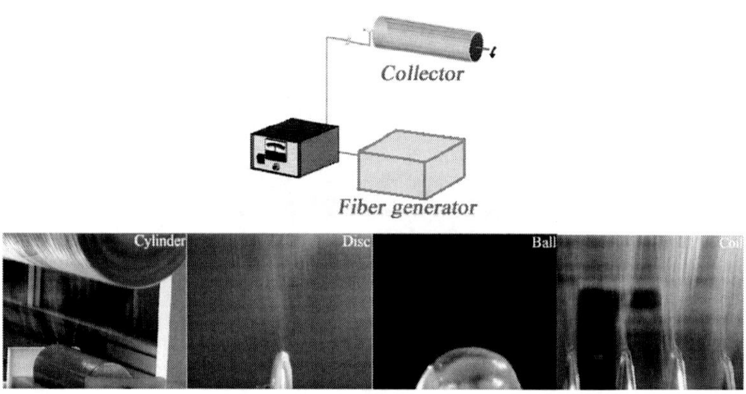

Figure 3.10 Schematic illustrations of needleless electrospinning setup, and cylinder, disc, ball, and spiral coil electrospinning processes [425–427].

Based on this understanding, a new needleless electrospinning system using a spiral coil wire as fiber generator was invented [426].

As shown in Fig. 3.10, when the applied voltage exceeded a critical value, numerous polymer jets were generated from the wire surface. It was also found that the spiral coil has higher fiber production rate than cylinder spinneret of the same dimension, and the average fiber diameter was finer with a narrower diameter distribution [427]. Details about these electrospinning setups and processes are provided in Chapters 4 and 5.

High throughput production of nanofibers in the name of "Tip-less Electrospinning" (TLES) has been demonstrated by Wu et al. [428], using a circular cylinder as the fiber generator. This is also an upward needleless electrospinning system using non-rotating spinneret. Experimental results showed that the yield of poly(ethylene oxide) nanofibers can be more than 260 times in weight compared to that of a single-jet electrospinning.

There are significant differences between the needle and upward needleless electrospinning processes. In the upward needleless electrospinning, Taylor cones are created on the surface of polymer solution. If the Taylor cone is stable, it will move together with the surface of the rotating roller and produces a solution jet under the strong electric field. Therefore, there must be strong inter-molecular interactions among polymer macromolecules in the solution to stabilize the Taylor cone given that the Taylor cone is stretched into a fine jet and deposited on the collector as solid fibers. Different to the conventional needle electrospinning in which the Taylor cone is generated and stabilized through constantly feeding polymer solution through the needle, the upward needleless electrospinning forms its Taylor cones by sucking up the solution covering the surrounding fiber generator [429]. It was observed that the base diameter of Taylor cone in cylinder electrospinning reduced from 1.2 to 0.3 mm in the beginning and the end of the electrospinning, respectively (initial polymer film thickness = 1 mm). If the film thickness is further reduced, no Taylor cone or nanofibers can be generated [428]. Therefore, the solution must have a suitable rheological property. Furthermore, a higher electric voltage is required to initiate the needleless electrospinning, because the Taylor cone is formed due to the wave fluctuation.

Chapter 4

Upward Needleless Electrospinning with Disc, Ball, and Cylinder Spinnerets

4.1 Cylinder Electrospinning

The cylinder has been proven to be a successful spinneret for needleless electrospinning, and cylinder-type electrospinning has already been commercialized by a company in the Czech Republic (Elmarco). Figure 4.1 shows the electrospinning process using a cylinder spinneret (cylinder diameter = 80 mm, fillet radius = 2 mm). Numerous solution jets were generated from the top surface of the cylinder and deposited on the collector, when the applied voltage was larger than a critical value.

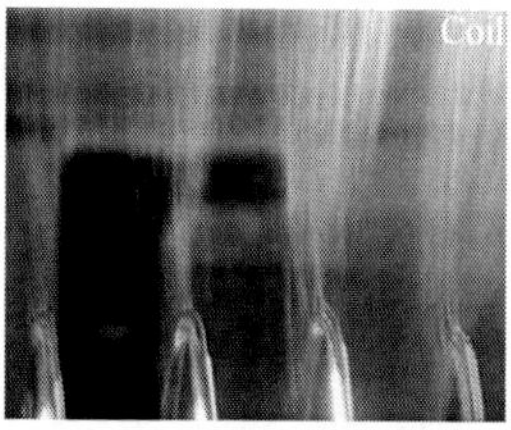

Figure 4.1 Illustration of a cylinder electrospinning process.

It was observed that the formation of jets/filaments was influenced by the spinneret rotating speed, the applied voltage, and the polymer concentration. The spinneret rotating speed

Needleless Electrospinning of Nanofibers: Technology and Applications
Tong Lin and Xungai Wang

ISBN 978-981-4316-35-4 (Hardcover), 978-981-4316-84-2 (eBook)
www.panstanford.com

affected the loading of the PVA solution onto the spinneret surface. When the rotating speed was less than 20 rpm, the coverage of polymer solution became uneven, and jets could not be generated continuously. Higher rotating speed, in the range of 40–50 rpm, enabled a thin layer of PVA solution to cover the spinneret surface evenly, resulting in continuous generation of jets/filaments from the spinneret surface. Further increasing the rotating speed could throw the polymer solution off the spinneret. Because of this, the spinneret speed was set at 40 rpm for all experiments mentioned in this chapter.

4.1.1 Effects of Applied Voltage and PVA Concentration

The electrospinning process using the cylinder spinneret showed a high dependence on the applied voltage. The critical applied voltage for generating jets from the cylinder spinneret was about 47 kV. Despite the high applied voltage, the jets were mainly generated from two end areas (about 2 cm wide) of the cylinder surface. A few jets were also observed to be produced on two sides of the cylinder. There was no jet/filament produced from the middle cylinder surface until the applied voltage was above 57 kV. Further increasing the applied voltage led to the generation of jets from the entire cylinder top surface, albeit with an uneven coverage, as illustrated in Fig. 4.1.

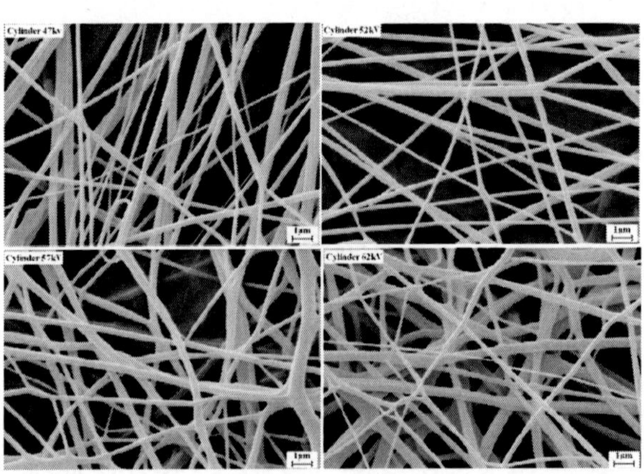

Figure 4.2 SEM images of PVA nanofibers electrospun by cylinder spinneret at different applied voltages (cylinder diameter = 80 mm, fillet radius = 2 mm, PVA concentration = 9.0 wt%, collecting distance = 13 cm). (At 47 and 52 kV the fibers were only generated from the cylinder ends.)

In addition to the electrospinning process, the fiber morphology was also affected by the applied voltage. The SEM images of nanofibers electrospun by the cylinder spinneret (cylinder diameter = 80 mm, fillet radius = 2 mm) from 9.0 wt% PVA solution under different applied voltages are shown in Fig. 4.2. The average fiber diameters calculated based on SEM images are given in Fig. 4.3a. For the cylinder electrospinning system, the electrospinning distance ranging from 11 to 15 cm was trialed. A shorter spinning distance led to wet fibers which merged into a polymer film on the collector, while a longer spinning distance resulted in stoppage of electrospinning because of the declined electric force. Nanofibers electrospun from 9 wt% PVA solution showed a bead-free fibrous structure. When the applied voltage was in the range of 47–62 kV, the average fiber diameters were very similar with a wide diameter distribution.

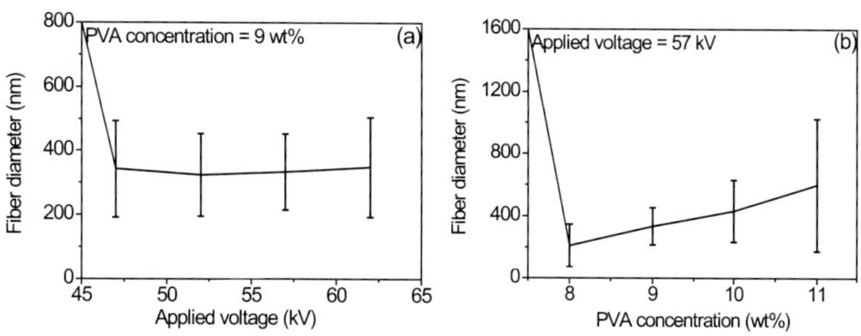

Figure 4.3 Dependencies of average fiber diameters on (a) applied voltage and (b) PVA concentration (collecting distance = 13 cm, cylinder diameter = 80 mm, fillet radius = 2 mm).

The cylinder electrospinning process was highly influenced by polymer concentration. When the PVA concentration was below 9.0 wt%, nanofibers were electrospun from the whole cylinder surface if the applied voltage was greater than 52 kV. However, for PVA solution with a concentration higher than 9.0 wt%, nanofibers were spun at the cylinder ends mainly, but not from the middle section of the cylinder. This was caused by the increased solution viscosity (larger than 1620 cP). In this case, the applied voltage had to be increased to spin fibers from the middle cylinder area. Also, higher PVA concentration led to coarser fibers produced from the cylinder ends. Fiber diameter was increased and the fiber diameter distribution became wider with the increase in PVA concentration (Fig. 4.3b).

The productivity of the cylinder electrospinning process was much dependent on the applied voltage and polymer concentration. As shown in Fig. 4.4, with an increase in the applied voltage, the production rate increased. Under the same applied voltage (57 kV), increasing the polymer concentration increased the productivity initially, but the productivity dropped when the PVA concentration was greater than 9 wt%. The reduced productivity at higher PVA concentrations can be attributed to the high solution viscosity, which restricted the formation of jet/filament.

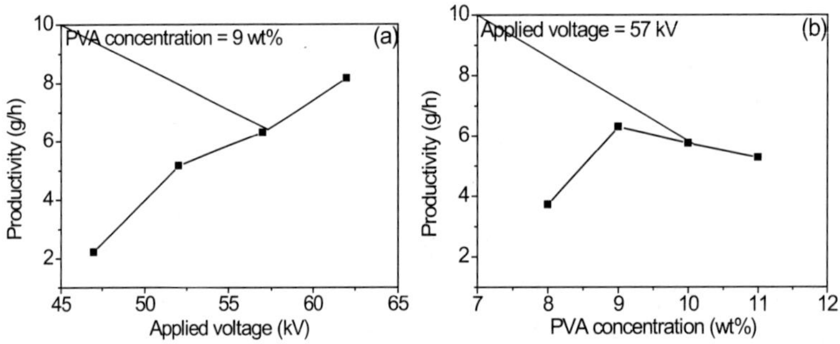

Figure 4.4 Dependencies of productivity on (a) applied voltage and (b) PVA concentration (collecting distance = 13 cm, cylinder diameter = 80 mm, fillet radius = 2 mm).

4.1.2 Effect of Cylinder End Shape

It has been noted that cylinder end shape influences the electrospinning process and fiber morphology. As listed in Table 4.1, when the cylinder spinneret had the same length and diameter, but varied in the end shape (fillet radius), electrospinning was only possible when the fillet radius was less than 20 mm. There were no obvious changes in the electrospinning processes, when the fillet radius varied between 2 and 10 mm.

All the electrospun nanofibers showed a bead-free fibrous structure, and there was little difference in the fiber morphology and fiber diameter (Fig. 4.5a).

The fillet radius of cylinders had little influences on average fiber diameter and diameter distribution. It is interesting to note that the cylinder with a 5 mm fillet radius gave higher productivity than cylinders with a higher or lower fillet radius (2 and 10 mm) (Fig. 4.5b).

Table 4.1 Influences of fillet radius and cylinder diameter on electrospinning process

Cylinder diameter 80 mm		Fillet radius 10 mm	
Fillet radius (mm)	**Electrospinning**	**Cylinder diameter (mm)**	**Electrospinning**
2	√	80	√
5	√	60	√
10	√	40	√
20	×		
30	×		

√—electrospinning success; ×—electrospinning failure; cylinder length—200 mm; applied voltage—57 kV; collecting distance—13 cm; PVA solution—9 wt%.

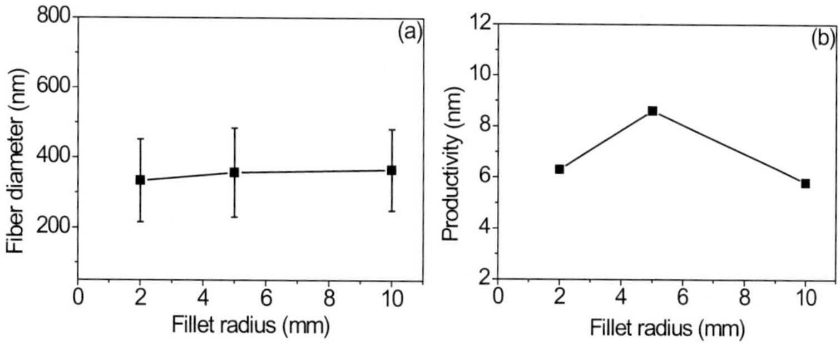

Figure 4.5 The influences of cylinder fillet radius on (a) fiber diameter and (b) productivity (cylinder diameter = 80 mm, applied voltage = 57 kV, collecting distance = 13 cm).

4.1.3 Effect of Cylinder Diameter

When the fillet radius was maintained at 10 mm, changes in the cylinder diameter had little effect on the electrospinning process (Table 4.1). Cylinders with a diameter smaller than 80 mm (e.g., 60 and 40 mm) could produce nanofibers as well. With smaller cylinders, the collector was lowered to keep the distance between the cylinder axis and the collector the same. The as-spun fibers also showed reasonably uniform fiber morphology. The cylinder diameter had little effect on the fiber diameter, but a smaller diameter cylinder resulted in greater fiber productivity (Fig. 4.6). With the small diameter cylinder, it was observed that the electrospinning process was improved in the middle cylinder area.

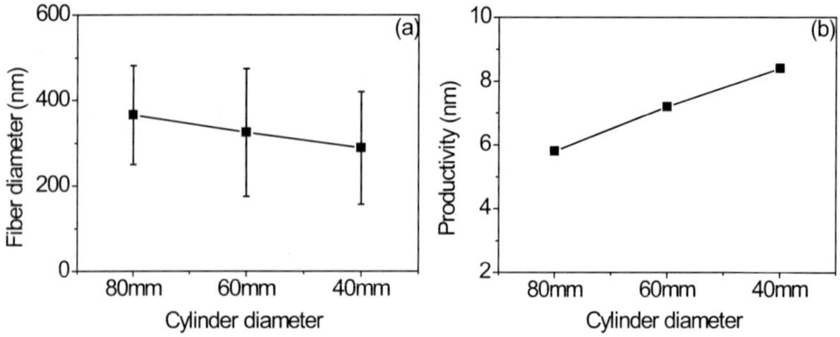

Figure 4.6 The influences of cylinder diameter on (a) fiber diameter and (b) productivity (fillet radius = 10 mm, applied voltage = 57 kV, collecting distance = 13 cm).

4.2 Ball Electrospinning

Although the cylinder spinneret has been demonstrated to have good electrospinning performance, the uneven electrospinning with the cylinder spinneret has been an issue. To solve this problem, a ball-shaped spinneret was used. As shown in Fig. 4.7, multiple jets are formed on the ball spinneret surface, indicating the success of electrospinning. The critical applied voltage to initiate an electrospinning process was 57 kV, which is higher than that of the cylinder spinneret of the same diameter. Once the electrospinning started, nanofibers were produced continuously. However, the nanofibers were mostly generated from the top of the ball, instead of the whole ball surface.

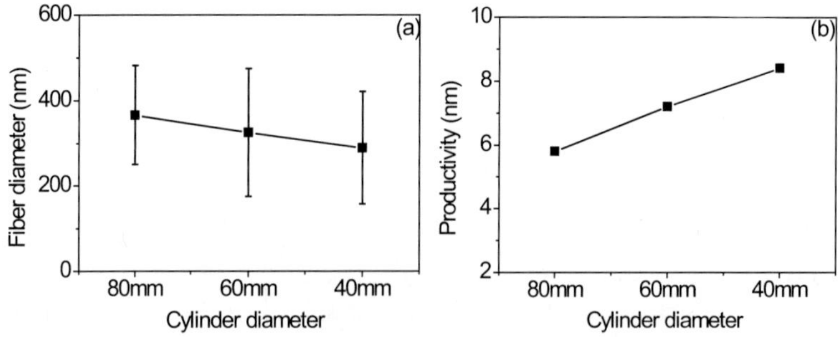

Figure 4.7 Illustration of a ball electrospinning process.

Figure 4.8a shows the morphology of nanofibers electrospun from 9 wt% PVA solution, which is similar to those electrospun from the cylinder spinneret under the same operating conditions (applied voltage = 57 kV, colleting distance = 13 cm). However, the ball spinneret had a much lower productivity (3.1 g/h) than the cylinder (6.3 g/h), possibly due to the smaller fiber generation area of the ball spinneret compared to that of the cylinder spinneret (Fig. 4.8b). There was very little difference between the average fiber diameters for these two systems.

Figure 4.8 (a) SEM image of PVA nanofibers produced by a ball spinneret (applied voltage = 57 kV, collecting distance = 13 cm, PVA concentration = 9 wt%); (b) the comparison of fiber diameter and productivity between the ball spinneret and cylinder spinneret (fillet radius = 2 mm).

4.3 Disc Electrospinning

Electrospinning using a disc spinneret is illustrated in Fig. 4.9. When the applied voltage was higher than a critical voltage, enormous jets were generated mainly from the circumambient area of the disc. The disc electrospinning process was very stable and showed little dependence on the environment.

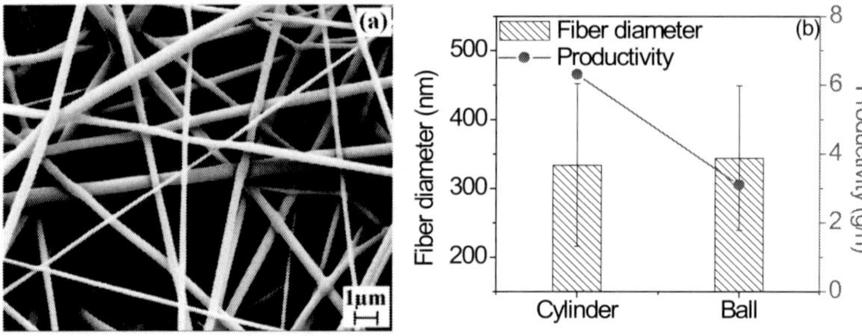

Figure 4.9 Illustration of a disc electrospinning process.

4.3.1 Effects of Applied Voltage and PVA Concentration

Similar to cylinder electrospinning, the disc electrospinning was affected by applied voltage also. When the applied voltage was lower than 42 kV, no fibers were produced on the disc spinneret surface. When the applied voltage was higher than 42 kV, nanofibers were generated continuously from the spinneret surface, and changing the applied voltage had little influence on the electrospinning process, until the voltage reached 62 kV, above which the working current in the power supply was too high (generating a large quantity of PVA nanofibers) to allow normal operation of the power supply.

Figure 4.10 SEM images of PVA nanofibers electrospun by disc spinneret under different applied voltages (Disc thickness = 2 mm, PVA concentration = 9.0 wt%, collecting distance = 13 cm).

Figure 4.10 shows the morphologies of nanofibers electrospun from 9.0 wt% PVA solution under different applied voltages. They were all bead-free fibers. The average fiber diameters calculated based on SEM images are given in Fig. 4.11a. With an increase in the applied voltage from 47 to 62 kV, the average fiber diameter reduced from 340 to 194 nm, and the diameter distribution also became narrower. At the same applied voltage, the fibers electrospun from the disc spinneret were slightly finer than those from the cylinder and ball spinnerets, and the higher the applied voltage, the finer the fibers produced from the disc spinneret.

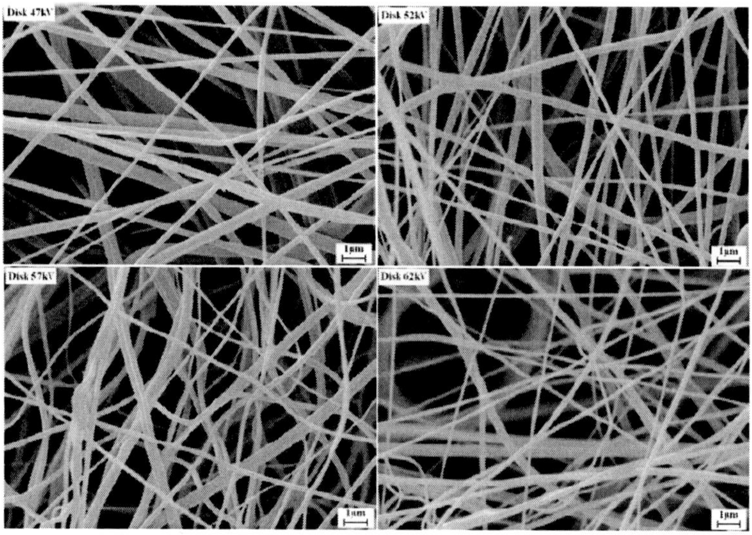

Figure 4.11 Dependences of fiber diameter on (a) applied voltage and (b) PVA concentration (collecting distance = 13 cm, disc thickness = 2 mm).

Polymer concentration was also an important factor affecting the electrospinning process and the fiber morphology. When the PVA concentration was in the range of 8.0–11.0 wt%, the solution could always be electrospun by the disc spinneret. The fibers electrospun from the disc spinneret increased in the fiber diameter with the increase in PVA concentration (Fig. 4.11b).

It was noted that the electrospinning distance for the disc electrospinning system could be adjusted between 11 and 19 cm, which is larger than that of cylinder spinneret. A shorter spinning distance led to wet fibers which merged into a polymer film on the collector, while a longer spinning distance resulted in stoppage of electrospinning because of the declined electric force.

With the collecting distance and PVA concentration kept constant, the production rate for disc electrospinning system increased steadily and reached about 8.2 g/h at 62 kV, which was about 3 times the value (2.6 g/h) at 42 kV (Fig. 4.12).

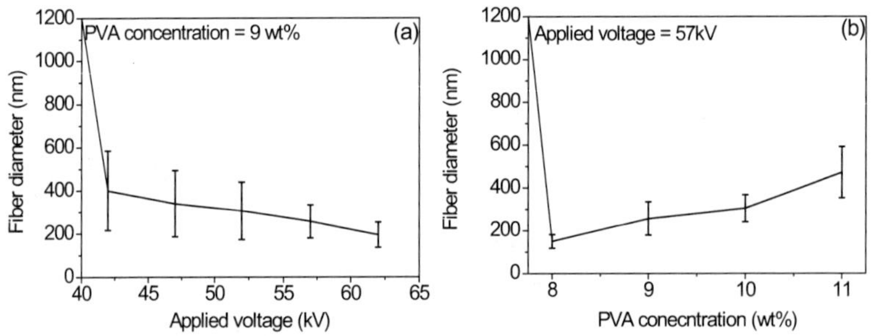

Figure 4.12 Dependencies of productivity on (a) different applied voltages and (b) different PVA concentrations (collecting distance = 13 cm, disc thickness = 2 mm).

Under the same applied voltage (57 kV), with an increase in the polymer concentration from 8 to 11 wt%, the productivity of disc electrospinning increased constantly, which was larger than that of the cylinder spinneret.

4.3.2 Effect of Disc Thickness

The thickness of disc also affected the electrospinning process. Discs with four different thicknesses (2, 5, 10, and 20 mm) and fillet radii (1, 2.5, 5, and 10 mm, respectively) were examined. Uniform PVA nanofibers were produced from all these disc spinnerets. PVA nanofibers from the 2 mm disc spinneret had an average diameter of about 250 nm (Fig. 4.13a), which reached about 320 nm for 20 mm disc spinneret. The average fiber diameter showed an increasing trend with increasing disc thickness. Also the productivity for the 2 mm disc spinneret had the largest value, about 6.2 g/h. The productivity showed a slightly decreasing trend with increased disc thickness (Fig. 4.13b).

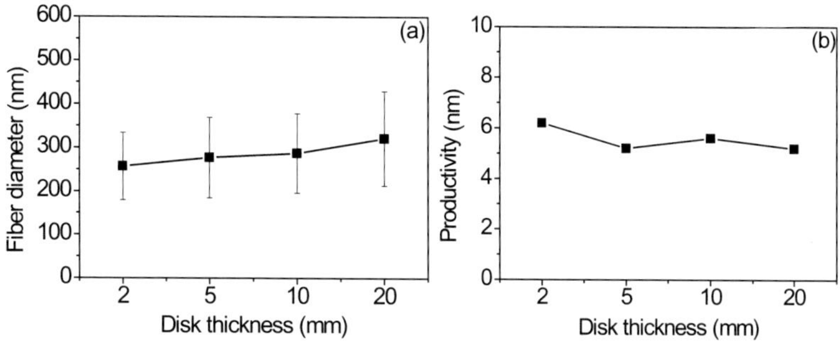

Figure 4.13 Influences of disc thickness on (a) fiber diameter and (b) fiber productivity (applied voltage = 57 kV, collecting distance = 13 cm, disc diameter = 80 mm).

4.3.3 Comparison

Comparisons of fiber diameters and productivities among cylinder, ball, and disc spinnerets are given in Fig. 4.14 showing that the disc spinneret has the best electrospinning performance. Disc spinneret produced PVA nanofibers with the smallest fiber diameter and narrowest diameter distribution. Also disc spinneret had a similar productivity to cylinder spinneret, which was much larger than that of the ball spinneret.

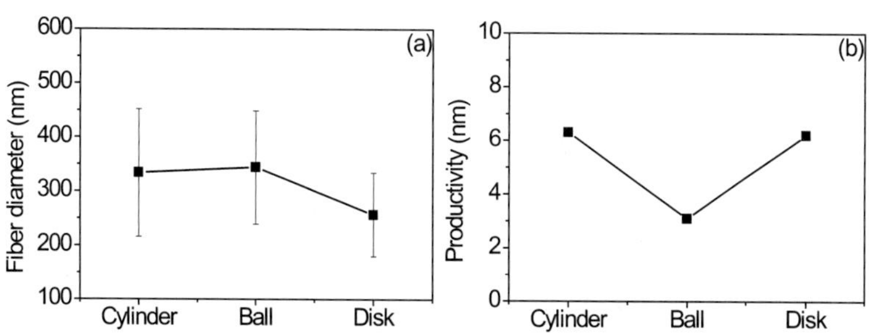

Figure 4.14 Comparisons among cylinder, ball, and disc spinnerets in (a) fiber diameter and (b) fiber productivity (applied voltage = 57 kV, collecting distance = 13 cm, PVA concentration = 9 wt%) (cylinder diameter = 80 mm, fillet radius = 2 mm, ball diameter = 80 mm, disc diameter = 80 mm, disc thickness = 2 mm).

Table 4.2 compares the diameter of electrospun PVA nanofibers and the productivity using different setups. The disc electrospinning can produce nanofibers with the smallest fiber diameter from almost all the PVA solutions. It should be noted that the needle electrospinning system works at a lower voltage and produces more uniform PVA nanofibers.

Table 4.2 Comparisons between needle and needleless systems

Concentration (wt%)	Fiber diameter (nm)			Production rate (g/h)		
	Needle[a]	Cylinder[b]	Disc[b]	Needle[a]	Cylinder[b]	Disc[b]
8	215	210	150	0.016	3.7	5.6
9	287	334	257	0.018	6.3	6.2
10	383	433	305	0.02	5.8	7
11	521	600	472	0.022	5.3	7.5

[a]Applied voltage = 20 kV, collecting distance = 13 cm, aluminum foil collector.
[b]Applied voltage = 57 kV, collecting distance = 13 cm, wire drum collector

Chapter 5

Upward Needleless Electrospinning with Ring and Coil Spinnerets

5.1 Ring Electrospinning

In the previous chapter, we have shown that cylinder and disc can be used to generate high electric field, which can be used for needleless electrospinning of nanofibers. Rings can be treated as hollow discs. They can also be used as fiber generators. Wire coils are similar to a group of interconnected rings. In this chapter, needleless electrospinning processes using rings and coils are discussed in detail.

5.1.1 Single Ring

A single ring was used as spinneret to generate nanofibers. As depicted in Fig. 5.1a, multiple jets were generated from the top of the ring when a high voltage was applied. The jet generation was found to be affected greatly by the experimental parameters such as applied voltage, collecting distance, and polymer concentration.

Needleless Electrospinning of Nanofibers: Technology and Applications
Tong Lin and Xungai Wang

ISBN 978-981-4316-35-4 (Hardcover), 978-981-4316-84-2 (eBook)
www.panstanford.com

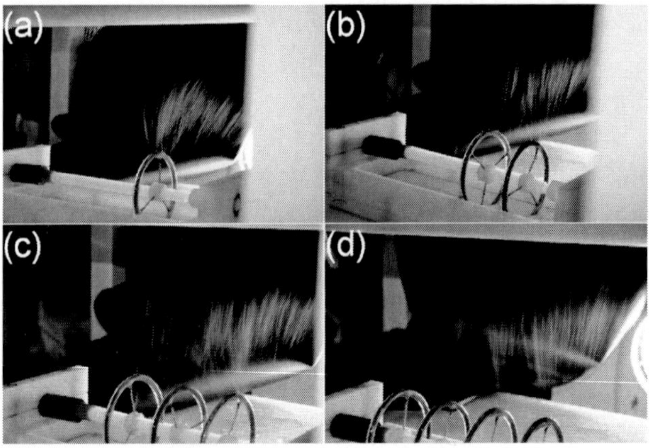

Figure 5.1 Photos of the electrospinning processes using ring spinnerets.

Figures 5.2 and 5.3 show the morphologies of the as-spun nanofibers. Some of the fibers are interconnected to each other owing to insufficient evaporation of solvent during the electrospinning process. Most of the fibers were very small in diameter, and the average fiber diameter was around 200 nm. Some coarse fibers were also observed when the applied voltage was low or the polymer concentration was higher than 9 wt%. This was due to either the low electrostatic force or the highly viscous nature of the polymer solution which limited stretching of the polymer jets.

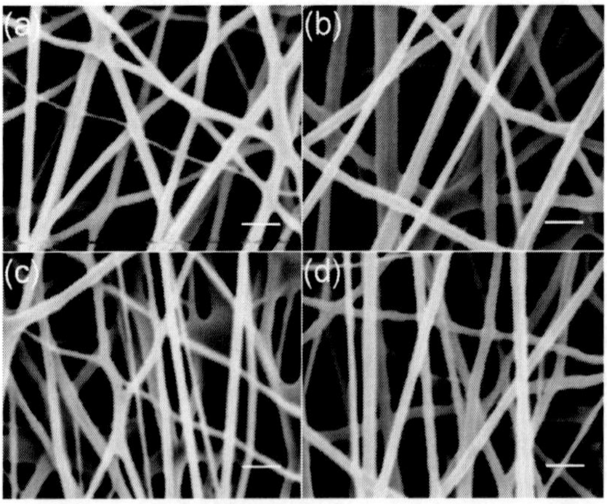

Figure 5.2 SEM images of nanofibers electrospun from a single ring coil at different applied voltages: (a) 50 kV, (b) 55 kV, (c) 60 kV, and (d) 65 kV. Bar = 1 μm.

The fiber diameter results are shown in Fig. 5.4. Fiber diameter decreased almost linearly with the increase of the applied voltage, and the diameter distribution was almost unchanged. On the other hand, the fiber diameter increased with the increase in the polymer concentration, and the diameter variation increased with the polymer concentration. The polymer solution having a high polymer concentration was difficult to electrospin. This was because of insufficient stretching resulting from the high solution viscosity. When the PVA concentration was 11 wt%, only a few jets were produced from the ring surface and the spinning process ran intermittently.

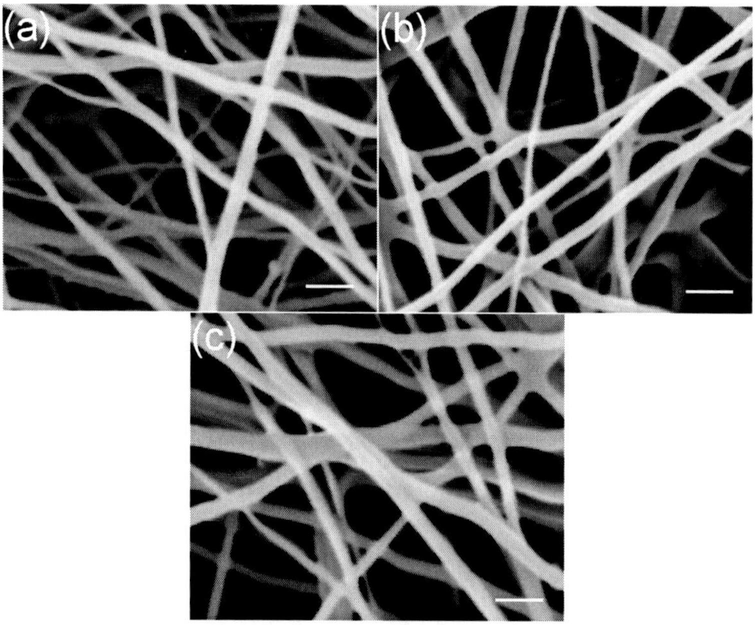

Figure 5.3 SEM images of nanofibers electrospun from a single ring coil with different PVA concentrations: (a) 8 wt%, (b) 9 wt%, and (c) 10 wt%. Bar = 1 μm.

The productivity of the ring electrospinning under different conditions is shown in Fig. 5.4. With the increase in applied voltage the productivity increased gradually. Increasing the polymer concentration, e.g., from 8 to 9 wt%, increased fiber productivity initially, and then the productivity decreased with further increase in polymer concentration.

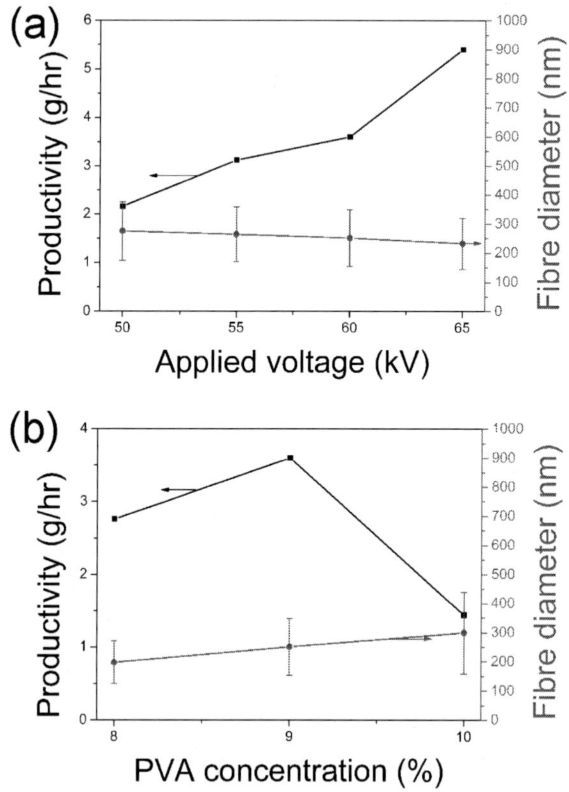

Figure 5.4 Productivity and fiber diameter of nanofibers electrospun from a single ring coil with different (a) applied voltages and (b) PVA concentrations.

5.1.2 Multiple-Ring Spinnerets

Increasing the number of rings is expected to increase fiber production rate. A multiple ring electrospinning setup is also shown in Fig. 5.1. Non-bead nanofibers were produced from the multiple ring system. These fibers showed a similar morphology with those that were electrospun from the single-ring electrospinning setup, except that more coarse fibers were observed, which became more evident with the increase in ring number.

The fiber diameter was calculated based on SEM images and the results are shown in Fig. 5.5. It is evident that with the increase in ring number, the average fiber diameter and diameter distribution increase. As expected the productivity increases swiftly with the increase of ring number (Fig. 5.5).

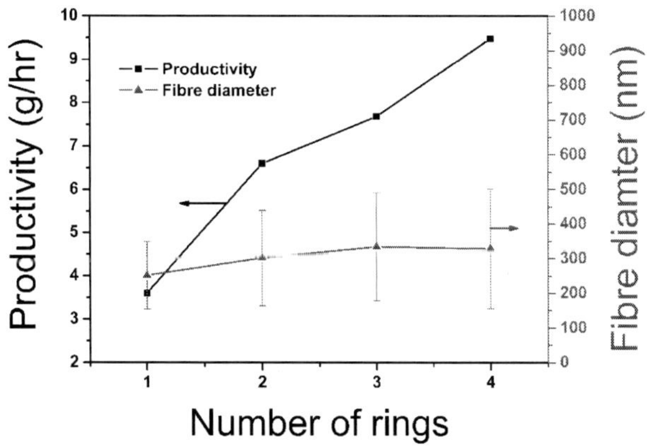

Figure 5.5 Productivity and diameter of nanofibers from ring electrospinning with different number of rings.

5.2 Wire Coil Electrospinning

5.2.1 Conical Coil Electrospinning

A conical wire coil was used to understand the needleless electrospinning process. Two reasons led us to using a conical wire coil, rather than a flat open surface, as spinneret. First, from geometry and fluidics, a conical wire coil is an open container which can hold more viscous fluid than a flat open surface. As a result of a slightly higher liquid pressure formed by the loaded fluid, the viscous fluid is easier to cover evenly underneath the surface of the cone. Second, a higher electric potential is normally formed on the outer surface of the cone. This enables the electrospinning of nanofibers only from the outer cone surface.

In electrospinning, when a viscous PVA solution was filled into the wire cone, it was retained by the conical coil without leaking out because of the surface tension and the viscoelastic nature of the polymer solution. When a high electric voltage was applied to the wire coil, the charged polymer solution moved down and covered the outer surface of the wire. A number of polymer jets were then generated mainly on the conical wire surface (Fig. 5.6a). Many polymer jets were drawn from the bottom of the spinneret, while others were from the wire on the conical surface. This was probably due to the distribution of the electrical filed which made the electrical force at the bottom of the nozzle quite high, and the gravity of the fluid which made the polymer gather at the bottom of the nozzle.

It was also found that the jets were generated from the areas that had sufficient polymer solution on the surface. Once the solution was exhausted temporarily in one area, the jet formation stopped, but restarted in other areas with replenished polymer solution, as shown in Fig. 5.6b–d. Figure 5.6e shows the locations of the jets generated. Polymer solution was filled in the spinneret and then little amount of the fluid was observed on the small slits between the wires due to the electrical force and the gravity of the polymer solution itself. The polymer jets were drawn out from either the top of the wires or the slit of the inter-wires because of the highly concentrated electric field.

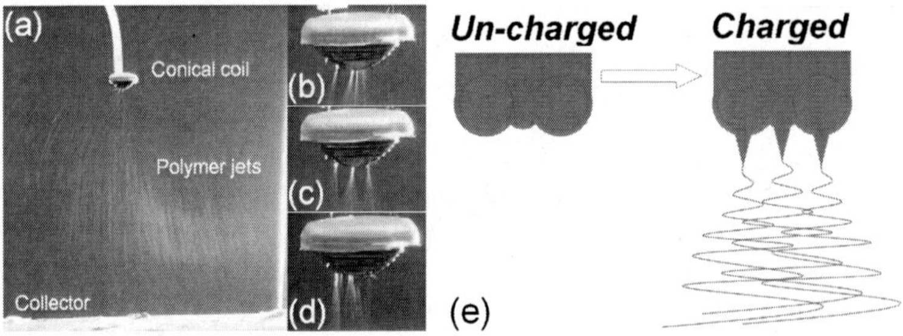

Figure 5.6 Needleless electrospinning using a conical wire coil. (a) Photo of the electrospinning process, (b–d) magnified view of the conical coil at different time, and (e) illustration of the jet formation on the coil surface.

For conventional needle electrospinning, the flow rate of the PVA solutions is always controlled in the range of 0.5–1.5 ml/h [430–432], to ensure no liquid drops off the nozzle tip, because a higher flow rate results in accumulating the extra liquid at the nozzle tip. For needleless electrospinning, numerous polymer jets were produced at the same time. The spinning rate was enhanced greatly compared to that of the needle electrospinning, and thus the productivity increased to a much higher level.

Figure 5.7a shows the spinning rate of the needleless electrospinning used in this study. It is obvious that the spinning rate varied greatly under different applied voltage and PVA concentration. The rate decreased with the increase in the PVA concentration. This was probably due to the increased solution viscosity making the liquid drops adhere at the slits thus hard to be drawn out. This was more evident when applied voltage was 70 kV, under which a large number of jets were produced. Besides, the spinning rate increased as the applied voltage increased. When the applied voltage was lower than 45 kV, the electrostatic force was weak and only a few jets were produced but not in a continuous manner. When the applied voltage was higher than 45 kV, more jets were observed because of the higher electric field formed on the nozzle surface which drew out more solution from the slits and more jets were produced accordingly.

Figure 5.7b shows the productivity under different applied voltages and polymer concentrations. The highest productivity is 2.75 g/h when the applied voltage was 70 kV and the polymer concentration was 9 wt%. Usually for needle electrospinning it takes a whole day to electrospin an A4 size nonwoven nanofibers mat. For the needleless electrospinning, using this conical wire coil setup, to produce a nanofiber mat of similar quality takes 1 h.

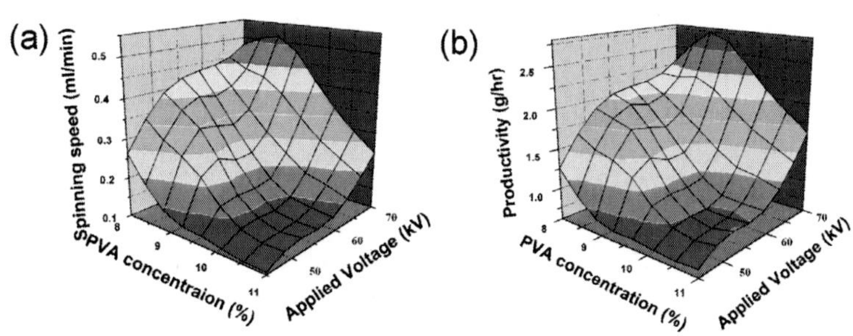

Figure 5.7 Spinning speed of the needleless electrospinning (a) and the converted productivity (b).

5.3 Effects of Experimental Parameters on Fiber Quality

5.3.1 PVA Concentration

The effect of PVA concentration on the spinning process was examined. The diameter of the electrospun PVA fibers was under 1 μm. When the PVA concentration was 8 wt%, most fibers were uniform. When the PVA concentration was increased to 11 wt%, the as-spun fibers became uneven and some coarse fibers with diameter around 1 μ were observed.

The diameter distribution of the nanofibers electrospun from different PVA concentrations was shown in Fig. 5.8a. When PVA concentration was 8 and 9 wt%, the diameter distribution of fiber diameter was around 200 nm. Much wider fiber diameter distribution was observed when the PVA concentration reached 11 wt%, and the fiber uniformity reduced.

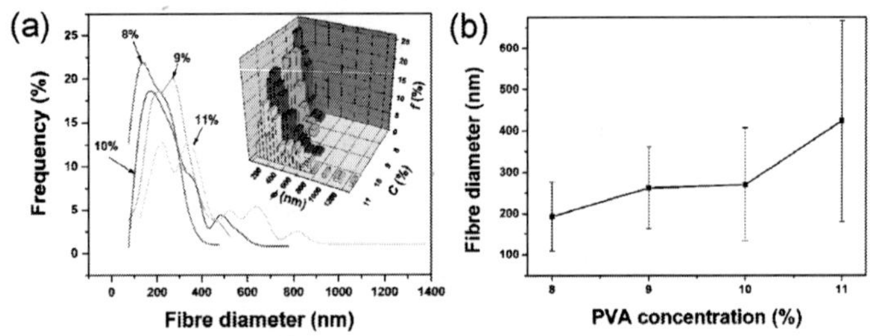

Figure 5.8 Diameter distribution (a) and average fiber diameter (b) of electrospun nanofibers with different PVA concentrations.

The average diameter and diameter distribution of the nanofibers are shown in Fig. 5.8b. The average diameter and diameter distribution increased gradually with the increasing PVA concentration. When PVA concentration changed from 10 to 11 wt%, the fiber diameter increased from 270 to 420 nm, and the diameter distribution expanded from 140 to 240 nm. Since the increase in PVA concentration led to increased solution viscosity, the higher viscoelastic property of the polymer solution weakened the fiber stretching during the electrospinning process. As a result, the average diameter increased accordingly. Besides, when the applied voltage was increased to 70 kV, polymer solution could not be retained by the wire coil structure, it was instead drawn out swiftly.

5.3.2 Applied Voltage

Figure 5.9a shows the diameter distribution of nanofibers electrospun at different applied voltages. All the distribution curves showed that the fiber diameter peaked at about 200 nm except for 45 kV, at which the main distribution peaked at around 300 nm. When applied voltage increased from 45 to 50 kV, the average fiber diameter decreased from 327 to 275 nm (Fig. 5.9b).

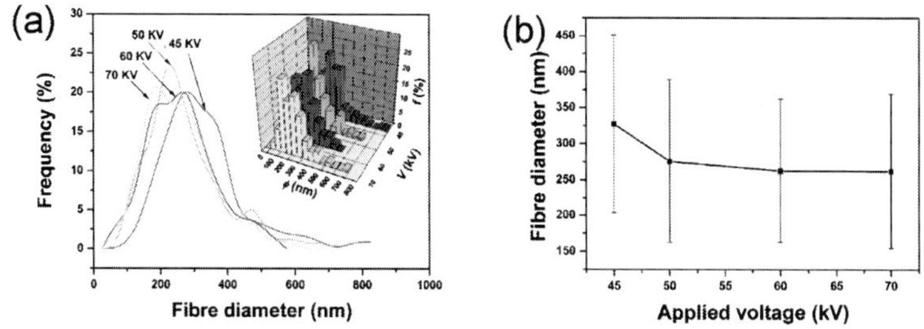

Figure 5.9 Diameter distribution (a) and average fiber diameter (b) of electrospun nanofibers with different applied voltages. PVA concentration 9%, collecting distance 15 cm.

5.3.3 Collecting Distance

Figure 5.10 shows the SEM images of PVA nanofibers electrospun under different collecting distances. All the nanofibers were thin and no large differences could be observed except image (Fig. 5.10a), in which some coarse fibers can be observed. When the collecting distance was 20 cm, the fibers traveled a long distance and solvent evaporated sufficiently. As a result the fibers were well isolated from each other. However, the fibers collected at a lower collecting distance stuck to each other and showed interconnected fibrous structure (Fig. 5.10b).

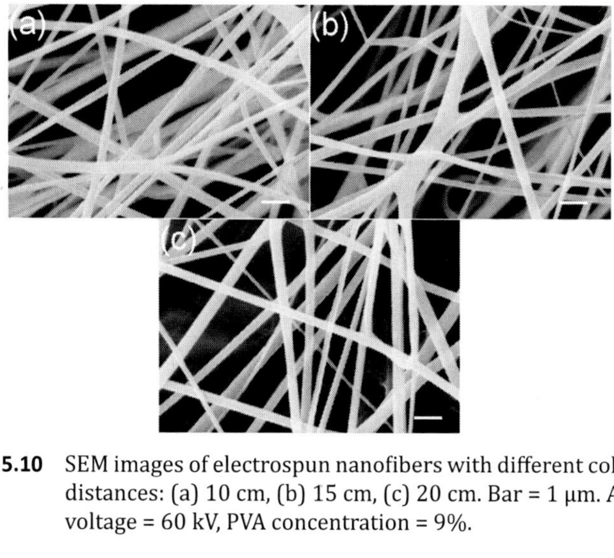

Figure 5.10 SEM images of electrospun nanofibers with different collecting distances: (a) 10 cm, (b) 15 cm, (c) 20 cm. Bar = 1 μm. Applied voltage = 60 kV, PVA concentration = 9%.

The diameter distribution is shown in Fig. 5.11a. The main peak of nanofiber diameter for the collecting distance of 20 cm was around 300 nm, which was greater than that of other collecting distances. Diameter and diameter distribution from the SEM images are shown in Fig. 5.11b.

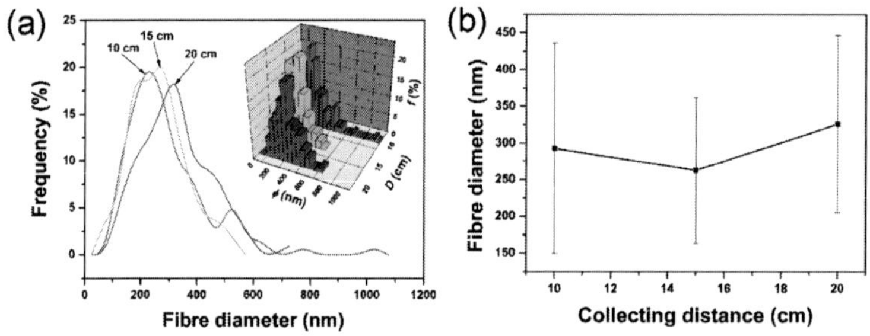

Figure 5.11 Diameter distribution (a) and mean diameter with deviation (b) of electrospun nanofibers with different collecting distance.

When the collecting distance was less than 10 cm, the polymer jets were insufficiently stretched, the fibers showed larger diameter and diameter distribution. Increasing the collecting distance could improve the attenuating effect on nanofibers. However, with the increase in the collecting distance, the electrical field strength decreases gradually which affects the fiber drawing. When the collecting distance was set at 20 cm, fewer jets were observed compared with that under lower collecting distances, and some PVA solution accumulated in the gap between the wires of the spinneret, larger solution drops were drawn out, and coarser fibers were produced. As a result, fibers collected with a collecting distance 20 cm produced coarser larger fibers with wider diameter distribution.

5.4 Tubular Wire Coil Electrospinning

Based on the above understanding of coil electrospinning, coil structure can be applied as the electric field concentrator and thus generating multiple polymer jets. Thinner nanofibers together with higher productivity were observed compared with needle electrospinning. However, the geometry of the conical wire coil limits the amount of polymer solution that can be used for mass production of nanofibers. By using ring/multiple rings structure, a large amount of polymer nanofibers can be produced at the same time; the only concern is the uneven deposition of nanofibers, leading to uneven thickness of the nanofiber web. A rotating tubular wire coil was thus used as spinneret to further enhance the fiber productivity and web uniformity. Multiple polymer jets were observed to generate on the top of each spiral coil. Coil shape and other experimental parameters were found to affect the fiber diameter and productivity.

5.4.1 Spinning Process

During electrospinning, the rotating tubular wire coil was partly immersed in the polymer solution. Because of the viscous properties of the solution, each coil was covered with a thin layer of polymer solution, and the solution layer on each spiral distributed evenly during the rotation. The coil together with the polymer solution was then connected to a high voltage so as to initiate needleless electrospinning. When the applied voltage was higher than 40 kV, polymer jets were first observed on the very top of the coil. With the increase of applied voltage, more and more polymer jets were observed and the jets were produced from a wider area of the coil.

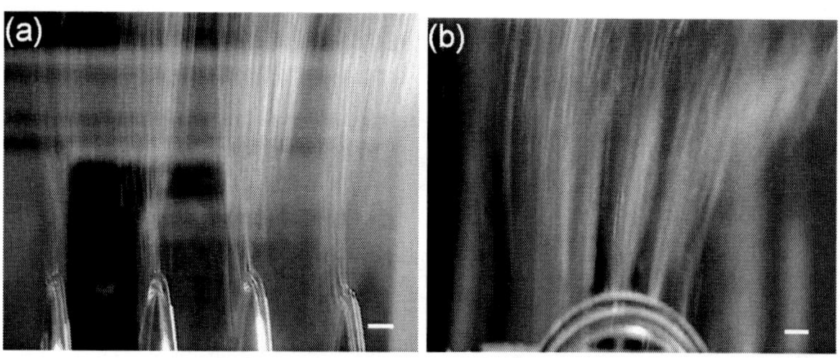

Figure 5.12 Photos of the tubular coil needleless electrospinning process from front view (a) and side view (b). Applied voltage = 60 kV, collecting distance = 13 cm, PVA concentration = 9 wt%. Bar = 1 cm.

Figure 5.12 shows the coil spinning process from both the front and side view. A large number of polymer jets were produced from each spiral during the spinning process. The side view reveals that the polymer jets were produced from the top arc (around 90 degrees) of each spiral. Unlike the conical wire coil described earlier, the tubular coil allows continuous loading of polymer solution from a solution bath during coil rotation, so that the electrospinning process can be easily scaled up for continuous and mass production of nanofibers.

5.4.2 Parameters Affecting Fiber Morphology

Figure 5.13 shows the SEM images of the as-spun PVA nanofibers/ microfibers from different PVA concentrations. Beads-on-string structure was formed when the PVA concentration was lower than 8 wt% (Fig. 5.13a). Similar to the conventional needle electrospinning, the formation of beaded fibers was attributed to an insufficient entanglement of polymer chains in the solution. With the increase of PVA concentration, the solution viscosity increased and the fiber diameter increased greatly as shown in the SEM images in (Fig. 5.13b–d). Electrospun fibers from polymer solutions with PVA concentration within 8–11 wt% showed uniform fiber morphology.

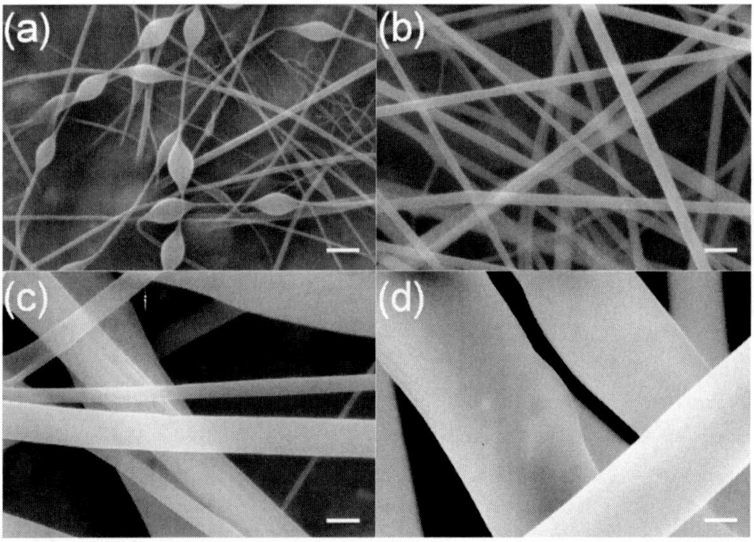

Figure 5.13 SEM images of electrospun nanofibers with different PVA concentrations: (a) 6 wt%, (b) 9 wt%, (c) 12 wt%, and (d) 15 wt%. Applied voltage = 60 kV, collecting distance = 13 cm. Bar = 1 μm.

For needle electrospinning, when PVA concentration was higher than 12 wt%, blocking of the needle happened frequently because of the high viscosity of the solution. In this case, a much higher applied voltage had to be used to increase the stretching force. However, "corona" discharge may happen if applied voltage is higher than 24 kV. In coil electrospinning, up to 70 kV can be applied and thus polymer solution with PVA concentration up to 15 wt% spinnable and microfibers can be produced, as shown in Fig. 5.13d. It should be pointed out that when the polymer concentration was higher than 15 wt%, fibers were electrospun intermittently, due to the highly viscous nature of the solution.

Operation parameters, such as applied voltage and collecting distance, also affected the fiber morphology. When the applied voltage was lower than 45 kV, many coarse fibers were formed. This was due to insufficient stretching under a weak electric field. With the increase in applied voltage, electric field intensity was strengthened gradually, less coarse fibers were formed, and fibers tended to be more uniform. Further increasing the applied voltage did not change fiber morphology much. It was also found that, even with a very high applied voltage, no apparent "corona discharge" was observed during the electrospinning process.

5.5 Effects of Coil Shape on Needleless Electrospinning

For needleless electrospinning, the spinneret length, inter-coil distance, coil diameter, and wire diameter were set at 16 cm, 4 cm, 8 cm, and 2 mm, respectively. The applied voltage, collecting distance, and PVA concentration were 60 kV, 13 cm, and 9 wt%, respectively.

5.5.1 Spinneret Length

Figure 5.14b shows the electric intensity of the tubular coil with different coil length *D*. The intensity of the middle spirals decreased with the increase in the *D*, but that of the side spirals remained almost the same. More polymer jets can be generated under a stronger electric field, and the fiber productivity of a single spiral increased with the decrease of *D*, as shown in Fig. 5.14a. The fiber diameter and the diameter distribution were almost the same. This is probably because the change in the electric field intensity was not enough to cause a large change in fiber diameter.

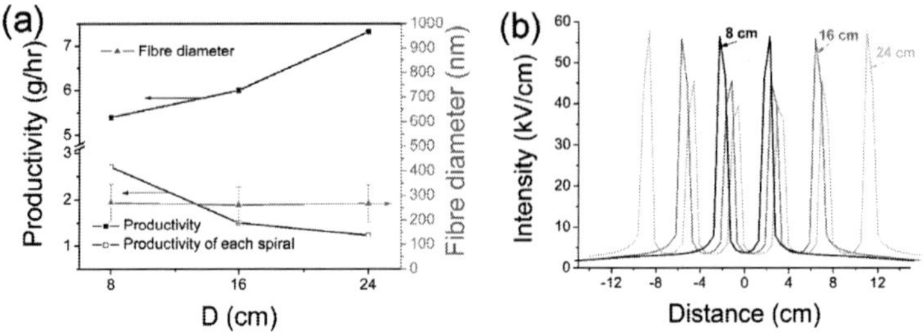

Figure 5.14 Effect of coil length on fiber diameter, productivity (a) and electric field intensity (b).

It is obvious that with the increase in D, the electric field intensity of the middle spirals decreased quickly. When the D is large enough, and the same applied voltage is used, the middle spirals may generate very weak electric field, and the whole coil will behave like a "cylinder". If the intensity is lower than a critical value, no jets will be produced from the middle coil surface, and the productivity decreases accordingly. In experimental, it was very hard to operate the electrospinning when the *D* is larger than 24 cm. Therefore, a higher voltage may have to be applied when a long coil nozzle is used.

5.5.2 Spiral Distance

When the spiral distance *d* (pitch) decreased from 8 to 2 cm, the productivity on each spiral decreased gradually since the electric field intensity decreased evidently (as shown in Fig. 5.15). It seems that the spirals interfered with each other when the pitch decreased (e.g., 2 cm). Very strong electric field was formed on the side spirals of the coil but much weaker field in the center, thus the fiber diameter together with diameter distribution increased (Fig. 5.15a).

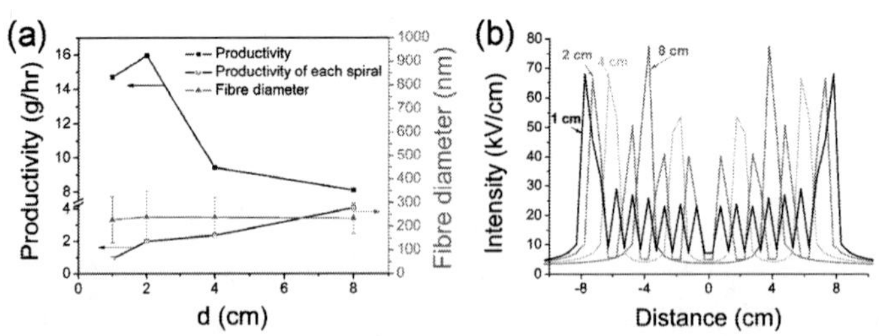

Figure 5.15 Effect of coil distance on fiber diameter, productivity (a) and electric field intensity (b).

When *d* was 1 cm, a strong electric field was formed on both sides of the coil (Fig. 5.15b), and lower intensity was found in the center area of the coil (5.15b). The whole intensity curve of this coil was very similar to that of the cylinder spinneret. The majority of polymer jets were produced from both ends of the coil and very few fiber jets were drawn out from the middle area. The average fiber diameter was thus smaller and the diameter distribution was much less than when $d = 2$ cm. Since few jets were produced in the center area, the overall productivity decreased a little as seen from Fig. 5.15a.

It is evident that reducing spiral distance *d* increased the productivity. The electric field intensity of the coil resembled that of a cylinder when *d* was further reduced.

5.5.3 Coil Diameter

The electric field intensity increased with the increase of coil diameter ϕ, as shown in Fig. 5.16b. More fibers were produced under higher electric field intensity which contributed to the increase of productivity. Also a larger area could generate polymer jets when coil diameter ϕ increased, and the productivity increased as well. The change of coil diameter ϕ had little influence on the fiber diameter and diameter distribution.

The solution level in the bath also affected the electric field intensity and its distribution. This is because any increase in the solution level would affect the electric field intensity profile and decrease the electric field intensity of the coil. The effect of decreasing the coil diameter is similar to that of increasing the solution level. An extreme case, when the coil diameter ϕ is 2 cm, is shown in Fig. 5.16b.

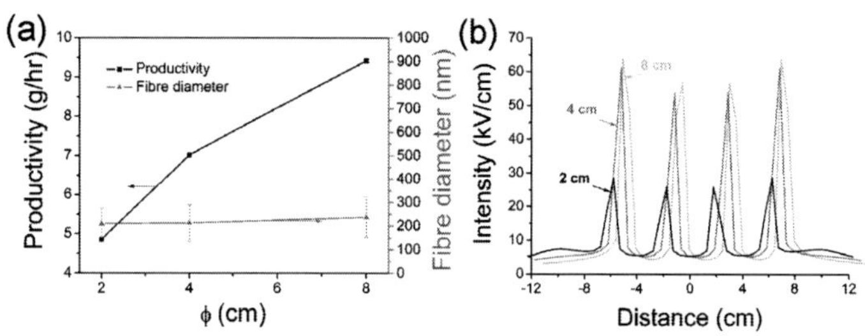

Figure 5.16 Effect of coil diameter on fiber diameter, productivity (a) and electric field intensity (b).

5.5.4 Wire Diameter of the Coil

The productivity of coil electrospinning reduced with the increase in wire diameter (Fig. 5.17b), because the electric field intensity generated by a finer wire was stronger (Fig. 5.17a). The higher electric strength in coils with smaller wire diameter produced more polymer jets on the wire surface resulting in increased productivity.

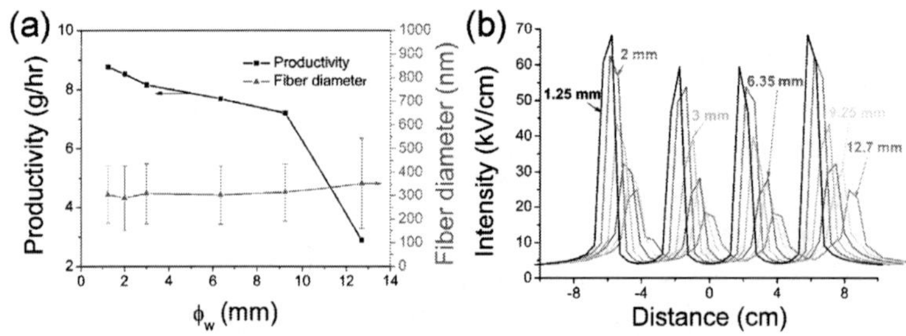

Figure 5.17 Effect of wire diameter on fiber diameter, productivity (a) and electric field intensity (b).

5.6 Effects of Experimental Parameters

5.6.1 Coil Rotating Speed

The coil rotating speed affected the loading of the PVA solution onto the coil surface. When the rotating speed was less than 20 rpm, uneven solution coverage became apparent, thus jets could not be generated continuously. Higher rotating speeds, in the range of 40–50 rpm, enabled the coil surface to be covered evenly with a thin layer of PVA solution, and thus jets/filaments were generated continuously. However, further increasing the rotating speed could throw polymer solution off the coil surface due to the higher centrifugal force.

5.6.1.1 Applied voltage

The electrospinning process is highly influenced by applied voltage. For tubular coil electrospinning, the minimum applied voltage required was around 40 kV, and the maximum applied voltage, without causing "corona discharge," was around 70 kV.

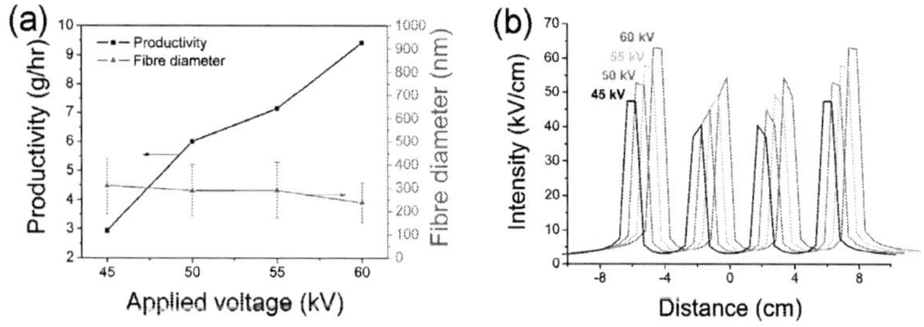

Figure 5.18 Effect of applied voltage on fiber diameter, productivity (a) and electric field intensity (b).

Figure 5.18a shows that increasing the applied voltage resulted in finer fibers with a narrower diameter distribution. This trend is similar to what happens in conventional needle electrospinning. It is because increasing the applied voltage strengthens the electric field between the coil surface and the fiber collector (Fig. 5.18b). When the applied voltage was 60 kV, the average fiber diameter was 237 ± 86 nm, which is much thinner than that of needle electrospun nanofibers. Figure 5.18a also shows that fiber productivity increased with the increase in the applied voltage.

5.6.2 Collecting Distance

The electric field intensity increased with the decrease of collecting distance, as shown in Fig. 5.19. The average fiber diameter decreased with the increasing collecting distance. Since the electric field intensity increased, the stronger electrostatic force made the fibers thinner and more uniform, and fiber productivity was increased greatly as well. However, if the collecting distance was too short, for example less than 6 cm, "corona discharge" between the coil and collector may occur, which will extinguish the jets production and harm the power supply in some cases. Besides, the solvent may not evaporate sufficiently over a short distance during electrospinning, which can result in a polymer membrane other than a nanofibrous web.

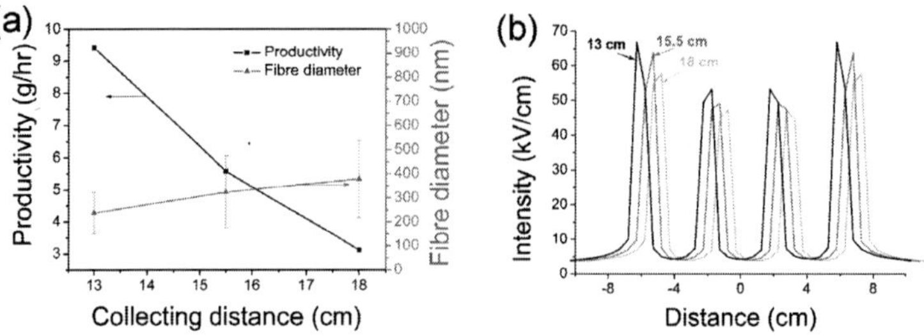

Figure 5.19 Effect of collecting distance on fiber diameter, productivity (a) and electric field intensity (b).

5.6.3 PVA Concentration

The increase in polymer concentration reduced the fiber productivity (Fig. 5.20). When polymer solution was higher than 12 wt%, very small numbers of polymer jets were produced. However, when the PVA concentration was decreased to 8 wt%, the fiber productivity dropped slightly. When the polymer concentration was very low (lower than 6 wt%), the solution was hard to pick up from the reservoir and thus no polymer jets were produced.

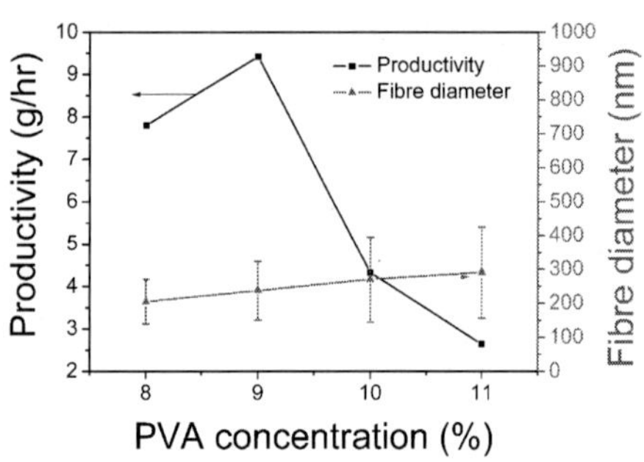

Figure 5.20 Effect of PVA concentration on fiber diameter and productivity.

5.7 Comparison

5.7.1 Needle and Coil Electrospinning

The typical morphologies of nanofibers electrospun from needleless and needle electrospinning are shown in Fig. 5.21. The nanofibers electrospun from both setups showed bead-free fibrous morphology. A number of fibers with very small diameter can be seen from coil electrospun fiber web (Fig. 5.21a). The fiber diameter of needleless electrospun fibers was much smaller than that from the needle electrospinning.

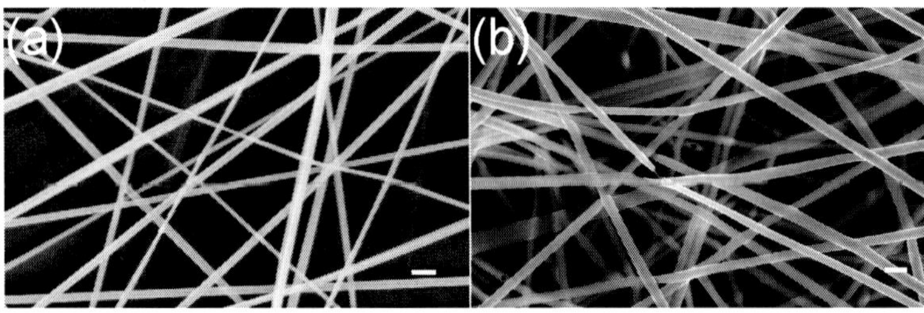

Figure 5.21 SEM images of nanofibers electrospun from (a) conical wire coil electrospinning (applied voltage = 60 kV) and (b) conventional needle electrospinning (applied voltage = 22 kV). PVA concentration = 9 wt%; collecting distance = 15 cm. Bar = 1 μm.

Figure 5.22 shows the flow rate of PVA solution being processed under different applied voltages for both the conical wire coil and a needle electrospinning system. For the wire coil system, the minimum voltage to generate jets was 45 kV, and the lowest voltage that led to the occurrence of corona discharge was 70 kV. Between the two voltage values nanofibers could be electrospun without any difficulty. With the increase in the applied voltage, the flow rate increased from 9.6 to 30.6 ml/h, and no liquid drops were observed to form from the spinneret.

The variation in applied voltage led to changes in the fiber fineness, the diameter of nanofiber decreased and then kept stable, also shown in Fig. 5.22. The fiber productivity for this electrospinning system can be estimated based on the volume rate. According to the flow rate data given in Fig. 5.22, the highest production rates for producing dry nanofibers were calculated to be 0.86 and 2.75g/h at 45 and 70 kV, respectively.

For comparison, a needle-based electrospinning setup (needle 21 Gauge, outer diameter = 0.82 mm; inner diameter = 0.51 mm) [433] was also used to electrospin nanofibers using the same polymer solution. The lowest voltages for initiating polymer jet and occurrence of "corona discharge" in the needle electrospinning were

8 and 25 kV, respectively. The range of operating voltage (8–24 kV) used for this electrospinning system was very similar to many other needle electrospinning processes. The fibers electrospun within this voltage range showed good fiber uniformity. A critical volume rate, the maximum flow rate of polymer solution not generating liquid drops under a given applied voltage during electrospinning, was used to understand the processing ability. As shown in Fig. 5.22, within the operating voltage range, the critical volume rate varied between 0.2 and 2.3 ml/h, and the volume rate value showed an increase with increased voltage applied. Based on the volume rate value, the highest rates of producing dry nanofibers from the needle electrospinning system were 0.018 and 0.207 g/h for the applied voltages at 8 and 24 kV, respectively.

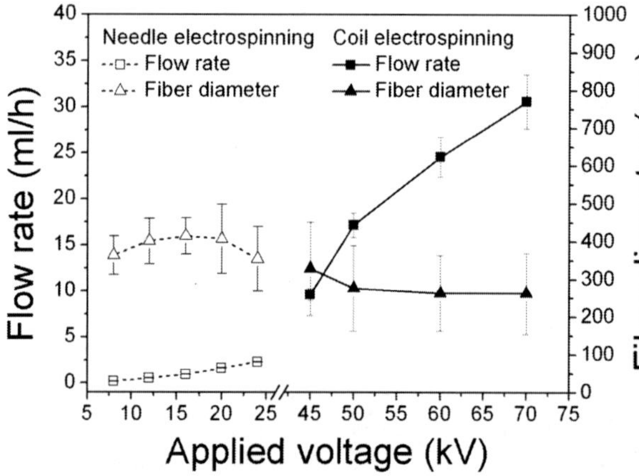

Figure 5.22 Dependence of volume rate and average fiber diameter on applied voltages of both needle and needleless electrospinning. PVA concentration = 9 wt%, collecting distance = 15 cm.

The average fiber diameter under the critical electrospinning condition is also shown in Fig. 5.22. With the increase in the applied voltage from 8 to 16 kV, the average fiber diameter increased gradually. Further increasing the applied voltage led to a slight reduction in the average fiber diameter. The variation of fiber diameter in the applied voltage range (8–24 kV) was between 353.4 ± 85 and 413 ± 48 nm.

Figure 5.23a shows the average fiber diameter of nanofibers produced from PVA solutions with different concentrations. The fiber diameter increased from 205.4 ± 65.1 to 290.8 ± 134.2 nm when the PVA concentration was increased from 8 to 11 wt%. For comparison, a needle-based electrospinning setup (needle 21 gauge, outer diameter = 0.82 mm, inner diameter = 0.51 mm) was used to electrospin nanofibers using the same polymer solutions with applied voltage 22 kV. Results are also shown in Fig. 5.29a and fiber diameter increases from 283.0 ± 59.4 to 685.6 ± 336.0 nm. It is apparent that nanofibers produced from the coil setup were much thinner than that from needle electrospinning setup, especially when the PVA concentration was higher than 9 wt%. Besides, the diameter distribution of needleless electrospinning was much narrower than that of needle electrospinning.

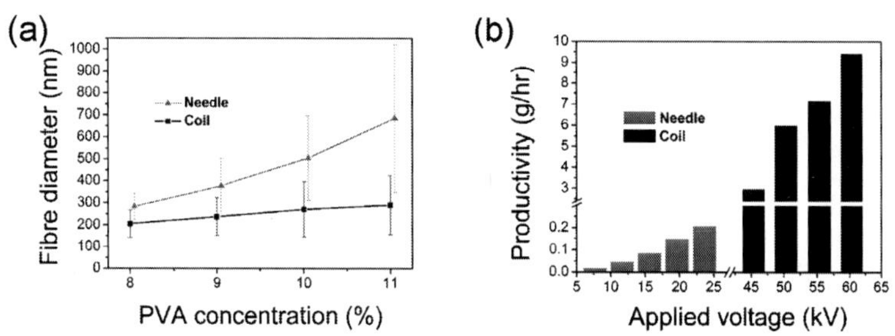

Figure 5.23 Comparison of fiber diameter (a) and productivity (b) between needle and tubular coil electrospinning.

Productivity of needleless electrospinning and needle electrospinning was also compared. Figure 5.23b shows the productivity of both needle electrospinning and needleless electrospinning. For the needle electrospinning, the productivity was increased from 0.01 to 0.21 g/h when applied voltage was increased from 8 to 24 kV. However, the productivity of coil needleless electrospinning increased from 2.94 to 9.42 g/h when the applied voltage was increased from 45 to 60 kV. Obviously the applied voltage is the dominant factor that determines the productivity of electrospinning. It is reasonable to expect that high productivity will be obtained if a higher voltage is applied. However, the applied voltage cannot be increased unlimitedly due to the "corona discharge".

Increasing the applied voltage in needleless electrospinning also increases the electrostatic force, which can draw more solution into jets at a given time; hence, contributing to the increased productivity. The production rate is affected by the spacing between the neighboring spirals, diameter of the coil, wire diameter, coil length, applied voltage, and external factors such as collector geometry which may affect the electric field that drives the electrospinning process.

5.7.2 Multiple Rings and Coils

Both ring and tubular coil can be used to produce nanofibers. Table 5.1 lists the experimental data from both tubular coil and ring coil. The size and length of both coils are the same. Four spirals and four rings were selected. It can be seen that the electric field intensity of the tubular coil spinneret was higher than that of rings. The nanofibers produced by tubular wire coil were much thinner and more uniform than that from the multiple rings.

Even though both coils gave very high fiber productivity, the collected nonwoven fiber mat from the tubular coil was more even than that from multiple rings. When the ring number was higher than two, gaps were formed on the collected nonwoven fiber mat. This is because polymer jets were repelled by other rings. One gap was formed when two rings were used, while two gaps from three rings, and so on. However, for tubular coil, all the spirals were connected and polymer jets were produced continuously along the axial direction of the coil, thus no gaps were formed and an even fiber mat was produced.

Table 5.1 Comparison between tubular wire coil and ring needleless electrospinning

	Spinnerets	
Items	**Rings**	**Tubular coil**
Spinneret length (cm)	16 cm	16 cm
Diameter (cm)	8 cm	8 cm
Coil distance (cm)	4 cm	4 cm
Electric field intensity (kV/cm)	62.3 (S), 52.2 (M)	66.5 (S), 53.8 (M)
Fiber diameter (nm)	328.6 ± 172.9	237.4 ± 86.1
Productivity (g/h)	9.48	9.42

Note: S = side coil, M = middle coil.

Theoretically, using multiple rings to perform needleless electrospinning on an industrial scale is more practical. However, considering the fiber morphology, tubular wire coil is better than a multiple ring system.

5.7.3 Comparisons of Different Nozzles

For cylinder, disc, ring, and coil, numerous polymer jets were produced from the top of all the fiber generators. The formation of jets was also greatly influenced by the spinneret rotating speed, the applied voltage, and the concentration of the polymer solution. The rotating speed of the spinneret affected the loading of the polymer solution on the surface of the spinnerets. A low rotating speed (less than 20 rpm) may lead to uneven solution coverage on spinneret surface and the jets could not be generated continuously. On the other hand, when the rotating speed was higher than 50 rpm, polymer solution was observed to be thrown off the spinneret surface by centrifugal force. A proper rotating speed, between 40 and 50 rpm, enabled the spinneret surface to be covered with a thin layer of polymer solution evenly, which is good for generating nanofibers.

The applied voltage is a very important parameter affecting needleless electrospinning. For cylinder, the critical applied voltage to start the spinning process was around 47 kV, and polymer jets were just produced at each end of the cylinder and no jets were formed in the middle area of the cylinder until the applied voltage was higher than 57 kV. For disc, the critical applied voltage to generate the spinning process was 42 kV, further increasing the voltage resulted in more polymer jets. However, for ring and tubular coil, the critical applied voltage to drive the spinning process was around 40 kV. From cylinder to coil, a strong electric field can be more easily formed on the spinneret top and much stronger electric field can be obtained under the same experimental conditions. It is obvious that among the four spinnerets examined, the coil structure is the most efficient nozzle for the mass production of nanofibers.

PVA concentration affected the formation of polymer jets as well. When PVA concentration was lower than 8 wt%, very few polymer jets were observed for all the four spinnerets and the spinning process was discontinuous. In most cases, no polymer jet was produced. When the concentration was higher than 11 wt%, owing to the highly viscous nature of the polymer solution, very few jets were produced and sometimes the spinning process stopped. The suitable range of PVA concentration for needleless electrospinning was found to be between 8 and 11 wt%.

All the fibers showed a nonwoven fibrous structure. In comparison with cylinder, finer fibers were observed in disc electrospun fibers. By using ring and tubular coil spinnerets, finer nanofibers were produced as shown in Fig. 5.24a,b. Since the electric field intensity of ring and tubular coil is much higher than that of the cylinder and the disc, polymer jets produced from ring and tubular coils receive higher electrostatic force which makes the fiber thinner.

Average fiber diameter and diameter distribution of the electrospun nanofibers from all four needleless spinnerets under different applied voltages and polymer concentrations are shown in Fig. 5.24.

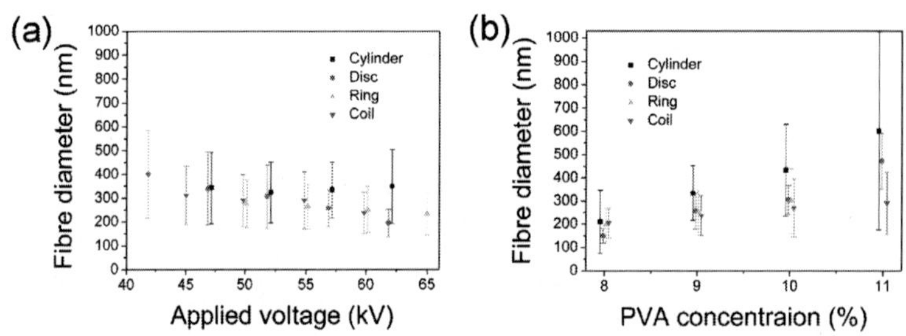

Figure 5.24 Experimental results of the fiber diameter of the nanofibers electrospun from different spinnerets: (a) different applied voltage (PVA concentration = 9 wt% and collecting distance = 13 cm); (b) different PVA concentration (collecting distance = 13 cm, applied voltage = 57 kV for cylinder and disc and 60 kV for ring and tubular coil).

For all the fiber generators, the fiber diameter decreased with the increase in applied voltage or the decrease in the PVA concentration. However, with the increase in applied voltage, the fiber diameter from cylinder nozzle did not change much and a high diameter distribution was also observed (Fig. 5.24a). For the other three spinnerets, not only the average fiber diameter but also the diameter distribution decreased with the increase in the applied voltage.

Electrospinning productivity of the four needleless spinnerets under different applied voltages and PVA concentrations is shown in Fig. 5.25. The productivity increased with the increase in the applied voltage for all the four spinnerets. Maximum productivity was observed at the PVA concentration 9 wt% for all the spinnerets except disc. This trend may be owing to either the strengthening of electric field or the change of the solution viscosity.

It is interesting to note that the productivity of cylinder and disc spinneret was almost the same, even though the disc was much smaller than the cylinder. Using disc spinneret could be much simpler way to produce thinner and uniform nanofibers on a large scale. Productivity of ring setup was less than disc as shown in Fig. 5.25. This is due to different spinnable area of the spinnerets. Even though the disc and ring have the same thickness and diameter, the width of the polymer solution covered on disc was larger than that on the ring. Besides, polymer jets may also be produced from the fringe area of the disc in addition to the top edge. However, for ring spinneret, just the very top area of the wire can produce polymer jets.

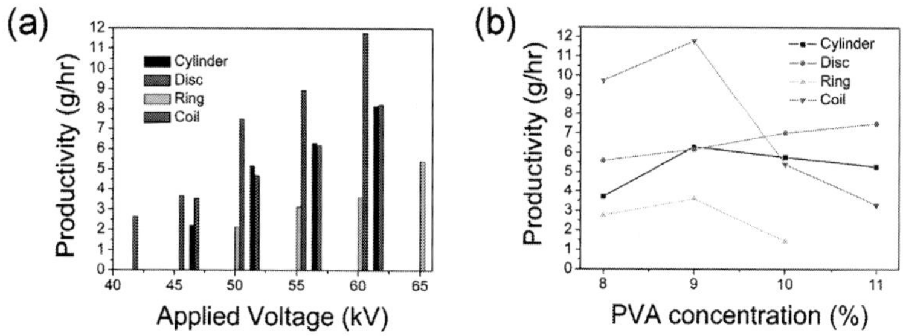

Figure 5.25 Experimental results of the productivity of nanofibers electrospun from different spinnerets under (a) different applied voltages and (b) different PVA concentrations.

Spiral wire coil has the highest productivity compared to other shapes of fiber generators, this is owing to the larger fiber generating area of the coil and also a much concentrated electric field formed. Using a tubular wire coil nozzle to generate nanofibers will greatly improve nanofiber production efficiency.

Figure 5.26 shows the FTIR spectra of PVA powder and PVA nanofibers electrospun from a conventional needle electrospinning and a tubular wire coil needleless electrospinning. The main peaks on the curves were ticked out and their corresponding vibrations are illustrated in Table 5.2. All three spectra show similar vibration peaks. This suggests that there are similar chemical components amongst the samples tested. The WAXD patterns of PVA powder and PVA nanofibers from both needle and needleless electrospinning are shown in Fig. 5.26 (right). All three curves show a prominent (101) peak at about $2\theta = 19.4°$.

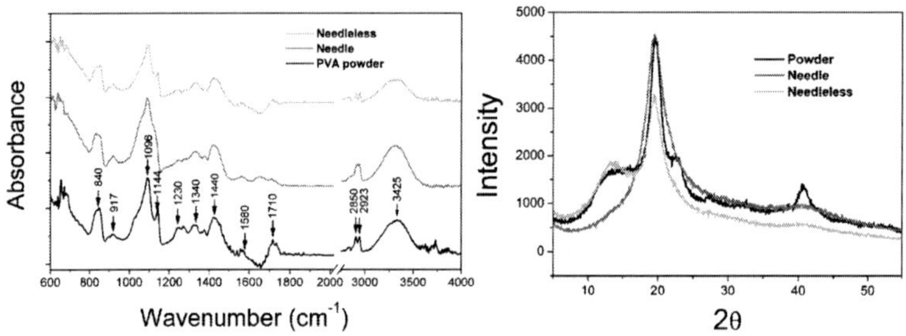

Figure 5.26 FTIR spectra and WXRD patterns of PVA powder, PVA nanofibers from needle and tubular wire coil needleless electrospinning.

Table 5.2 Characteristic vibration peaks of the FTIR spectra

Wave number of peak (cm^{-1})	Corresponding to
3425	hydroxyl group (O–H) stretching
2850, 2923	C–H alkyl stretching band
1710	C=O vibration
1440	C–H bend of CH_2 of PVA
1235–1340	C–H wagging and CH-OH bending vibration
1260	C–C stretching
1144	C–O–C bending vibration
1096	C–O stretching vibration
917	CH_2 rocking
840	CH rocking

Chapter 6

Electrical Field Analysis

In needle electrospinning, the polymer solution at the nozzle tip receives four main forces: surface tension force, electric force, viscoelastic force, and gravity, as shown in Fig. 6.1. Surface tension tends to reduce the liquid surface area. Induced by a high electric voltage, charge accumulation occurs on the liquid surface, giving rise to the electric force. At a low applied voltage, under the influence of electric force, the droplet is deformed to balance the forces received. With an increase in the applied voltage, the solution droplet evolves from a hemi-sphere to a cone shape (Taylor cone) and a high electric force is thus concentrated at the apex of the Taylor cone. When the electric field is above a critical value, the droplet at the cone tip overcomes its surface tension. Jet is formed and ejected into the electric field formed between the nozzle tip and collector.

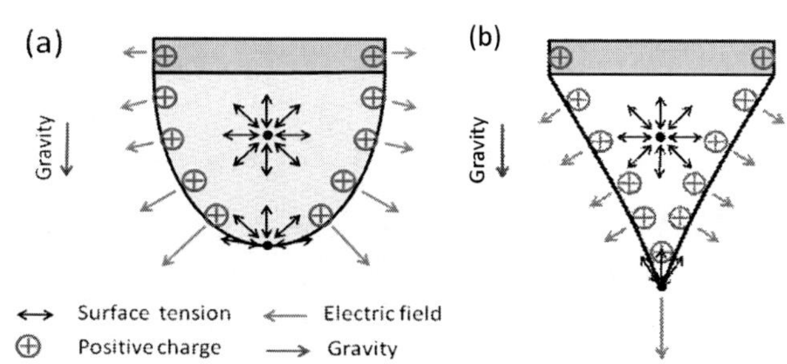

Figure 6.1 Force analysis of charged droplet at two different periods.

Needleless Electrospinning of Nanofibers: Technology and Applications
Tong Lin and Xungai Wang

ISBN 978-981-4316-35-4 (Hardcover), 978-981-4316-84-2 (eBook)
www.panstanford.com

The jet initiation is determined by the applied voltage. In needle electrospinning, the critical applied voltage for electrospinning was proposed in Eq. (6.1) [434]:

$$V_c^2 = 4\ln\left(\frac{2h}{R}\right)(1.3\pi R\gamma)(0.09) \tag{6.1}$$

This predication was based on a Zeleny's needle/point apparatus (Fig. 6.2), where h is the distance from the needle tip to the collector, R denotes the needle outer radius, and γ is the surface tension. The factor 0.09 is inserted to predict the voltage. V_c is the critical voltage.

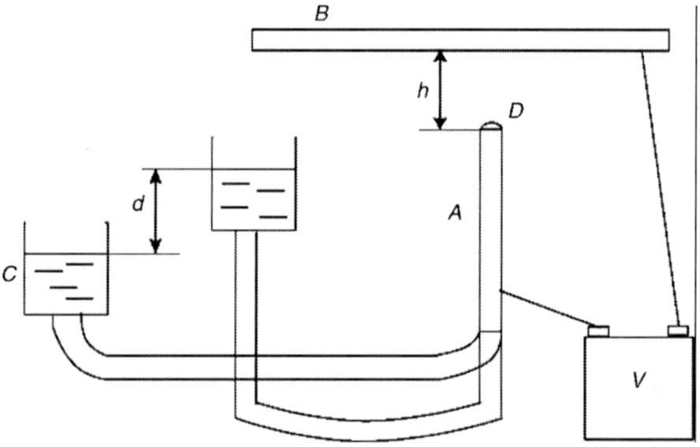

Figure 6.2 Zeleny's needle/point apparatus.

Ludas et al. [424] have explained the self-organization of jets happening on a free liquid surface in needleless electrospinning. The critical electric field intensity was proposed as:

$$E_c = \sqrt[4]{4\gamma\rho g/\varepsilon^2} \tag{6.2}$$

where ρ is liquid mass density, g is gravity acceleration, γ is surface tension. In either of the models above, electric force plays a crucial role in the jet initiation. In a static electric field, the relationship between voltage and electric field intensity can be expressed as:

$$E = -\Delta V \tag{6.3}$$

The calculated electric field for needle spinneret is shown in Fig. 6.3. The small diameter needle creates high intensity electric field, about 47 kV/cm at the needle tip, when the applied voltage is as low as 20 kV. From needle to the collector, the electric field intensity decays rapidly initially and then stabilizes at a relatively low level.

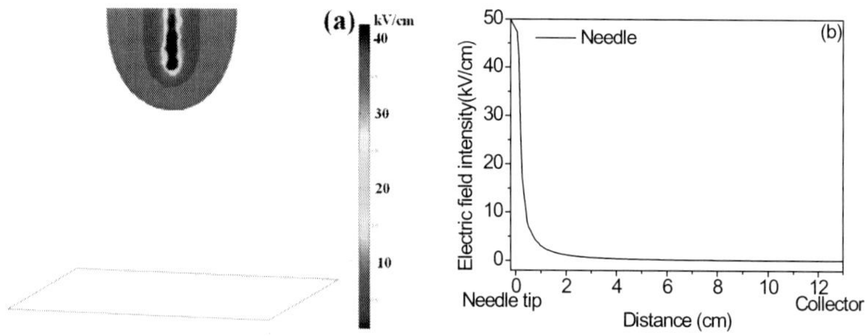

Figure 6.3 (a) Electric field isosurface of needle electrospinning, (b) electric field intensity from the needle tip to collector (applied voltage = 20 kV, collecting distance = 13 cm).

6.1 Cylinder Electrospinning

The electric field in needleless electrospinning setups is analyzed using the same method. The electric field intensity profiles of cylinder electrospinning setup are shown in Fig. 6.4. A high electric field is formed on the cylinder surface, and the electric field at the cylinder ends shows much higher intensity than that in the middle area. With increasing applied voltage, the increase in electric field intensity is mainly at the cylinder ends, and the electric field in the cylinder middle area changes little with the variation of applied voltage.

The rectangular shape below the cylinder indicates the polymer solution, which is held in a PTFE container. Although the corners of the solution container generate quite high intensity electric field, their effects on the electrospinning process is negligible in practice, because the container is made from a highly insulate PTFE material.

The electric field intensities of the cylinder spinneret under different applied voltages are obtained from Fig. 6.4. As shown in Fig. 6.5a, the front view of electric field intensity profile along the cylinder surface confirms that along the cylinder surface, electric field with a high intensity is formed at the cylinder ends, and levels off at a relatively low value in the middle surface area. From spinneret surface to the collector, the electric field at the cylinder ends decays rapidly in the initial 2 cm and then stabilizes. However, the electric field from the cylinder middle to the collector decays rather slowly (Fig. 6.5c).

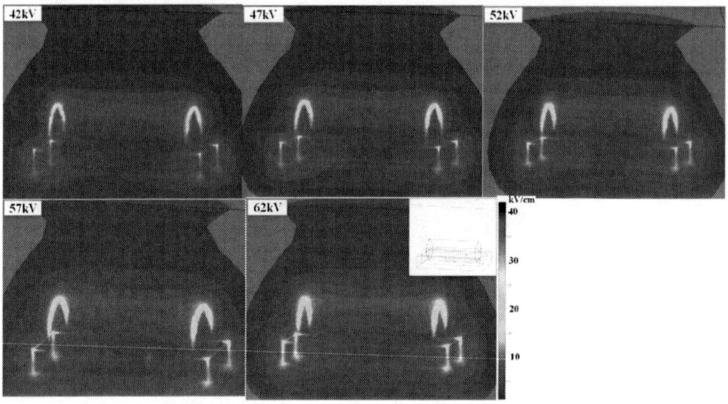

Figure 6.4 Electric field isosurfaces of cylinders under different voltages (cylinder diameter = 80 mm, fillet radius = 2 mm, collecting distance = 13 cm).

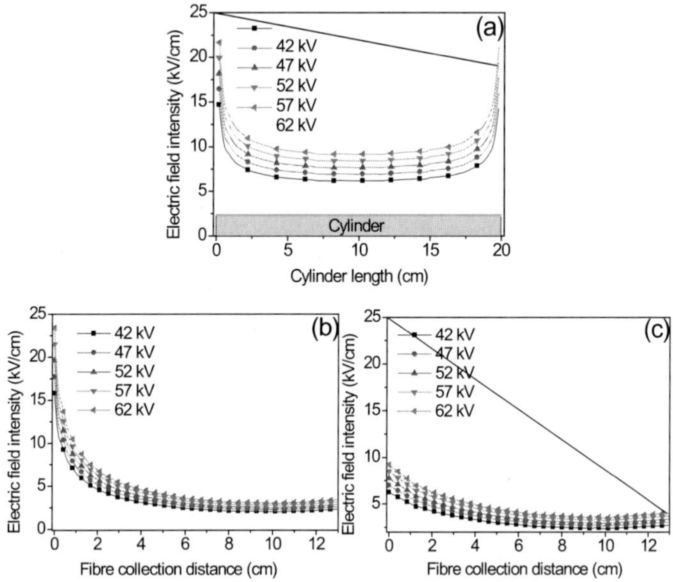

Figure 6.5 Electric field intensity of cylinder electrospinning setup under different applied voltages. (a) Front view of electric field intensity along the cylinder length, (b) electric field intensity from cylinder end to the collector electrode, (c) electric field intensity from cylinder middle to the collector electrode (cylinder diameter = 80 mm, fillet radius = 2 mm, collecting distance = 13 cm).

To understand how the cylinder end shape influences the electric field profile, the electric field intensity profiles of cylinder spinnerets with the same diameter, but different fillet radii are analyzed. As shown in Fig. 6.6, with an increase in the fillet radius, the electric field intensity at the cylinder ends reduces. However, the intensity in the cylinder middle area is little affected by the fillet radius.

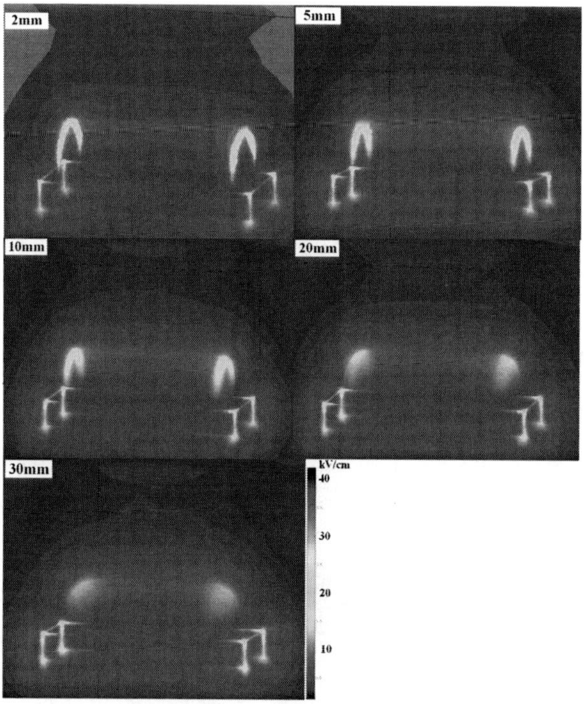

Figure 6.6 Electric field isosurfaces of cylinder spinnerets with different fillet radii (cylinder diameter = 80 mm, applied voltage = 57 kV, collecting distance = 13 cm).

Figure 6.7 shows the electric field intensity profiles on the cylinder and the influence of cylinder end dimension. At the cylinder ends, a larger fillet radius leads to reduced electric field intensity at the cylinder end, but the electric field intensity in the cylinder middle area remains nearly unchanged (Fig. 6.7a,b). There is very little change in the electric field profile from the middle section of the cylinder to the collector for the cylinders with different radii (Fig. 6.7c).

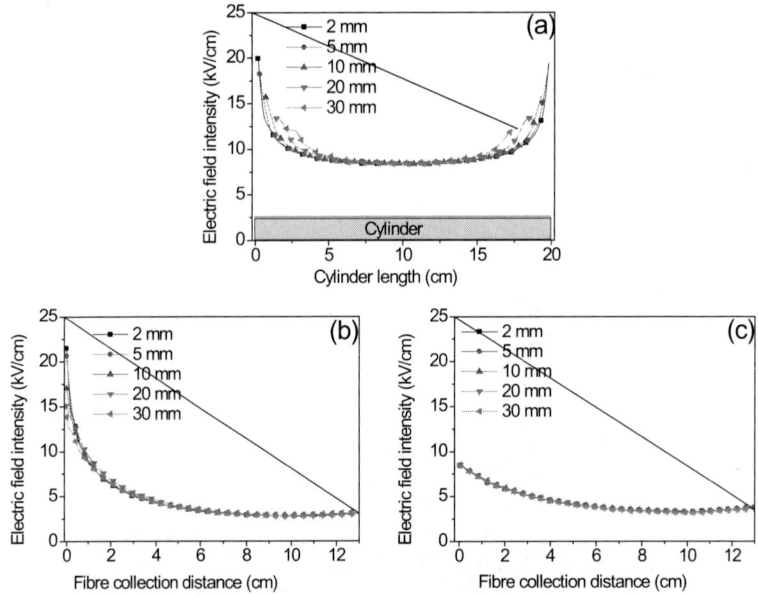

Figure 6.7 Electric field intensity of cylinder spinnerets with different fillet radii, (a) along the cylinder length, (b) from cylinder end to the collector, (c) from cylinder middle to the collector (cylinder diameter = 80 mm, applied voltage = 57 kV, collecting distance = 13 cm).

Figure 6.8 Electric field isosurfaces of cylinders with different diameters (fillet radius = 10 mm, applied voltage = 57 kV, collecting distance = 13 cm).

Besides cylinder end shape, cylinder diameter also influences the electric field intensity profile (Fig. 6.8). As shown in Fig. 6.9, along the spinneret length, the electric field intensity in cylinder middle area reduces with an increase in the cylinder diameter, but electric field intensity at the cylinder end change very little. It has been noted that the electric field intensity for the cylinder spinneret with a diameter of 60 mm shows a very similar profile to that of the spinneret with a diameter of 80 mm. Also, the electric field intensity from spinneret surface to the collector does not change much.

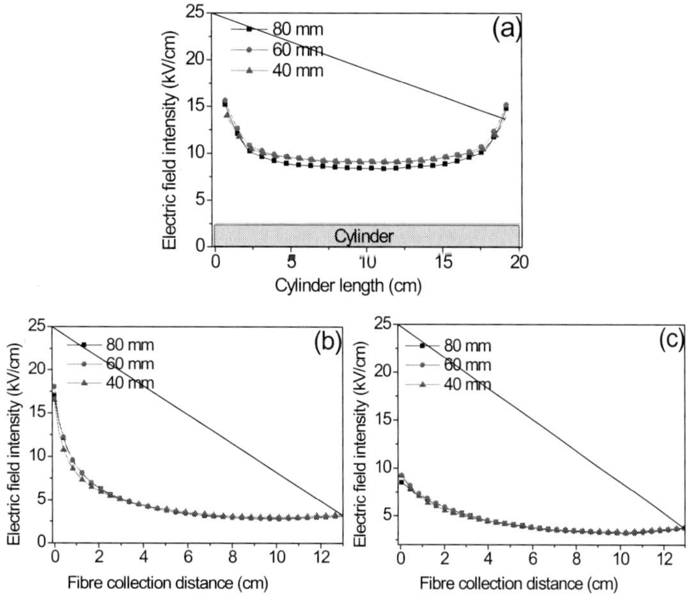

Figure 6.9 Electric field intensity of cylinders along electrospinning direction with different diameters (a) along the surface of cylinder, (b) from cylinder end to the collector, (c) from cylinder middle to the collector (fillet radius = 10 mm, applied voltage = 57 kV, collecting distance = 13 cm).

These results suggest that the cylinder dimensions (e.g., diameter, fillet radius, length) affect the electric field intensity along the spinneret surface, and that the electric field intensity varies significantly along the cylinder length. In addition, the decay in electric field intensity from the spinneret surface to the collector is faster at the cylinder ends than in the cylinder middle area.

6.2 Ball Electrospinning

Since cylinder spinneret shows uneven distribution in electric field intensity along the spinneret surface, a ball spinneret is expected to have more uniformly distributed electric field along the spinneret surface. The calculation results of the electric field profile for ball spinneret are shown in Figs. 6.10 and 6.11. High intensity electric field is mainly generated on the top of the ball spinneret, and the electric field shows a more uniform distribution than that generated by cylinder spinneret.

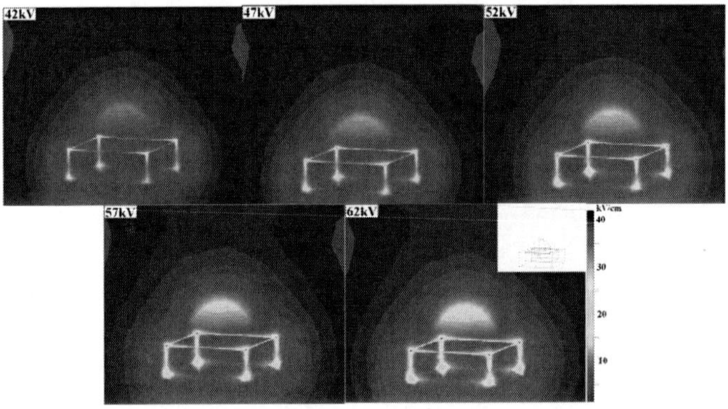

Figure 6.10 Electric field isosurfaces of ball spinnerets (ball diameter = 80 mm, applied voltage = 57 kV, collecting distance = 13 cm).

From ball spinneret surface to the collector, the electric field intensity decays at a slower rate than that at the cylinder ends but faster than in the cylinder middle. In comparison with the cylinder spinneret, one can see that the electric field intensity at the ball is only about 14 kV/cm, which is smaller than that of cylinder ends (21 kV/cm), but larger than that of cylinder middle (8.5 kV/cm) when the same applied voltage (57 kV) is employed.

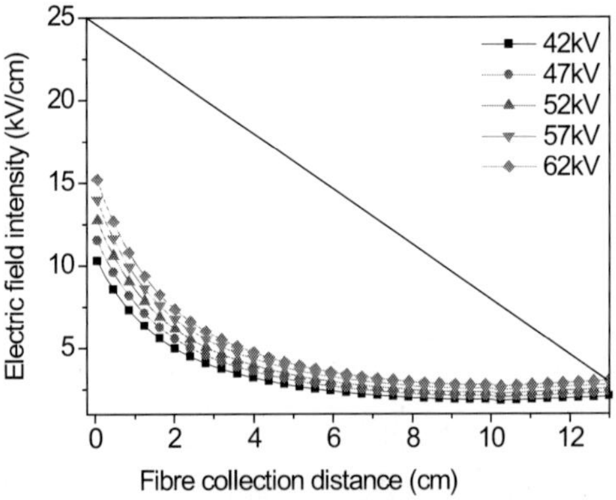

Figure 6.11 Electric filed intensity of ball spinnerets (ball diameter = 80 mm, applied voltage = 57 kV, collecting distance = 13 cm).

6.3 Disc Electrospinning

Figure 6.12 shows the calculated electric field intensity profiles of disc electrospinning setup. Very high intensity electric field is observed on the disc edge. Under higher applied voltage, the stronger electric field intensity appears over a larger disc edge area. In comparison with cylinder spinneret and ball spinneret, the disc spinneret forms much stronger electric field around the disc top circumambience. Although the highest electric field intensity is formed on the circumference of the disc, the electric field intensity on other parts is very low.

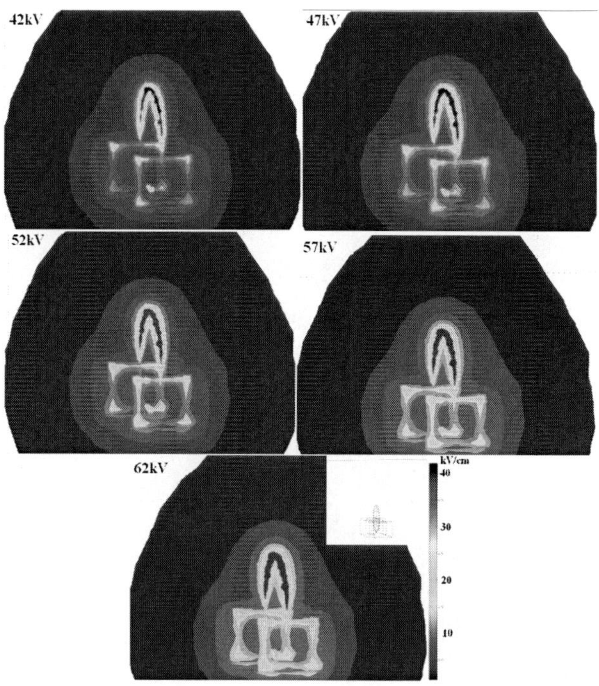

Figure 6.12 Electric field isosurfaces of disc spinnerets with different applied voltages (disc thickness = 2 mm, collecting distance = 13 cm).

At the same applied voltage and collecting distance, disc spinneret generates an electric field with a much higher intensity than the cylinder and ball spinnerets, and the electric field also decays more rapidly from the disc surface to the collector (Fig. 6.13).

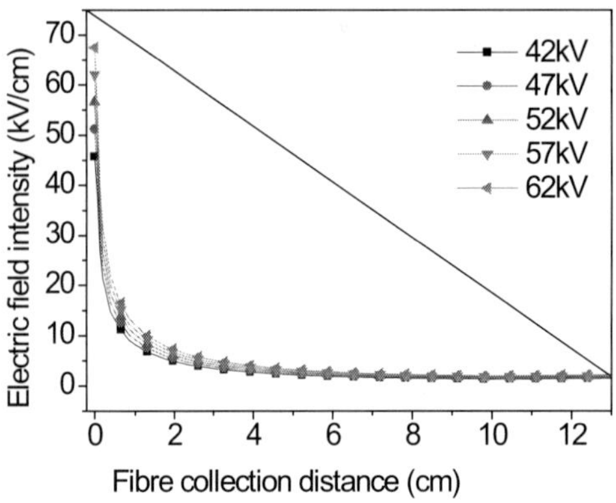

Figure 6.13 Electric field intensity along the electrospinning direction from the disc top to the collector (disc thickness = 2 mm, collecting distance = 13 cm).

As shown in Fig. 6.14, the generated electric field is affected by the disc thickness. With an increase in the disc thickness, the electric field intensity on the disc decreased, but the area with relatively high intensity increased.

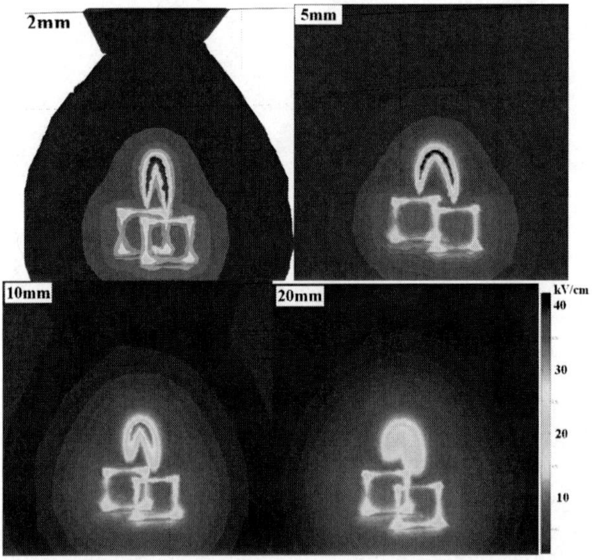

Figure 6.14 Electric field isosurfaces of disc spinnerets with different thicknesses (disc diameter = 80 mm, applied voltage = 57 kV, collecting distance = 13 cm).

From Fig. 6.15a, it is obvious that a thinner disc produced higher electric field intensity at the disc top. The electric field intensity is calculated from the disc tip to the collector for discs of different thicknesses (Fig. 6.15b). All the intensity curves show similar trends. The largest value is formed at the disc top, which decays rapidly away from the disc, and finally stabilizes near the collector. The thinner the disc, the stronger the electric field on the disc top.

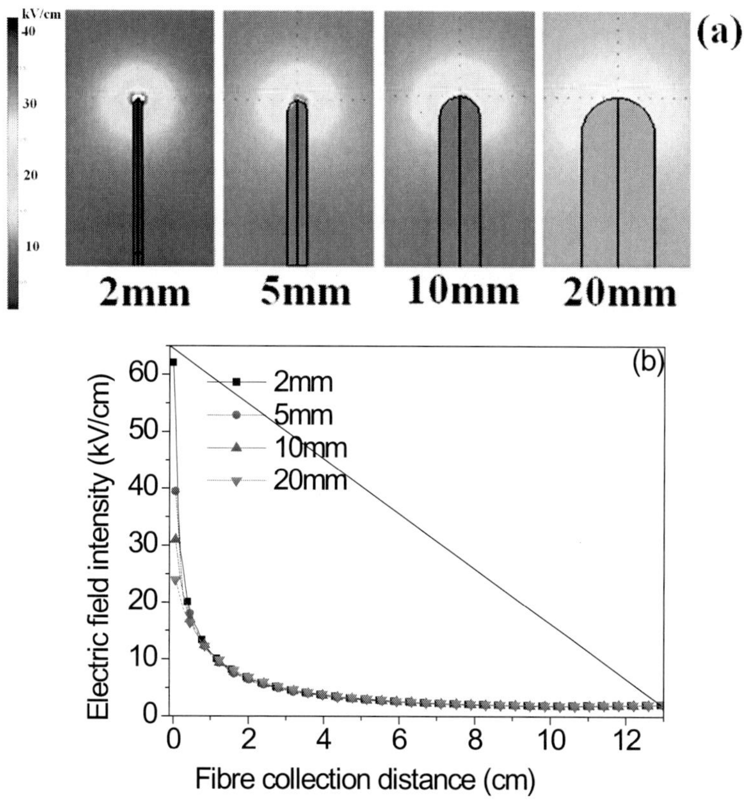

Figure 6.15 Electric field isosurfaces (a) and their intensity (b) of disc spinnerets with different thicknesses (disc diameter = 80 mm, applied voltage = 57 kV, collecting distance = 13 cm).

The electric field intensities for the three spinnerets are listed in Table 6.1. Under the same applied voltage, the electric field intensity generated by the disc is the highest amongst the spinnerets examined.

Table 6.1 Electric field intensity at the spinneret surface (kV/cm)

Applied voltage (kV)	Cylinder		Ball	Disc
	Ends	Middle		
42	19	6	10	46
47	21	7	12	51
52	23	7.7	13	57
57	25	8.5	14	62
62	27	9	15	67

Collecting distance = 13 cm for all applied voltages, diameter = 80 mm, cylinder fillet radius = 2 mm, disc thickness = 2 mm.

Figure 6.16 shows the front view of the electric field generated by the cylinder, ball, and disc spinnerets. It shows that the disc spinneret can produce the strongest electric field, and the electric field at the end and middle areas of cylinder are apparently different. The ball spinneret generates weak but even electric field on the top surface.

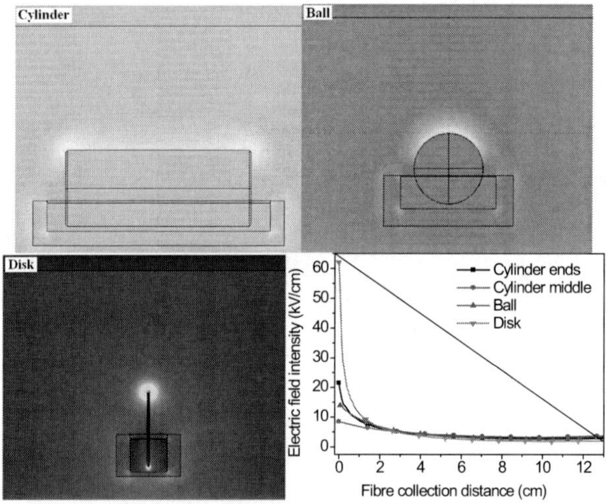

Figure 6.16 Electric field isosurfaces and intensity of the cylinder, ball, and disc spinnerets (collecting distance = 13 cm, applied voltage = 57 kV, diameter = 80 mm).

The essential requirements for the initiation of an electrospinning process are a high intensity electric field and a strong interaction between the electric field and polymer fluid. The electric force has been identified as the driving force of an electrospinning process. Although the driving force for electrospinning is the same, the electrospinning process could be influenced by the spinneret structure.

6.4 Shaft-Linked Multiple Discs

Multiple discs linked with a shaft are shown in Fig. 6.17. The diameter of the shaft is in the range of 0–40 mm. The inter-disc distance is between 0 to 40 mm.

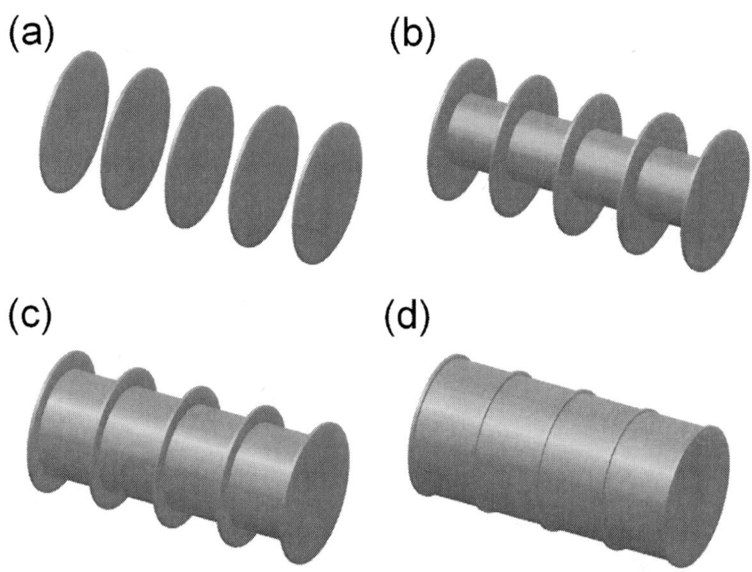

Figure 6.17 Schematics of multiple discs with different shaft diameters: (a) 0 mm, (b) 20 mm, (c) 30 mm, (d) 38 mm.

Figure 6.18 shows the electric field intensity profiles of multiple discs with different shaft diameters. High electric field is formed on the ends of the spinnerets. Strong electric field is formed on the top of each disc, as shown in Fig. 6.18a. With the increase of the shaft diameter, the electric field intensity on the disc surface becomes weaker. When the shaft diameter increases to 38 mm, the electric field intensity of the middle discs is almost the same as that on the shaft surface, which is almost the same as that of a cylinder.

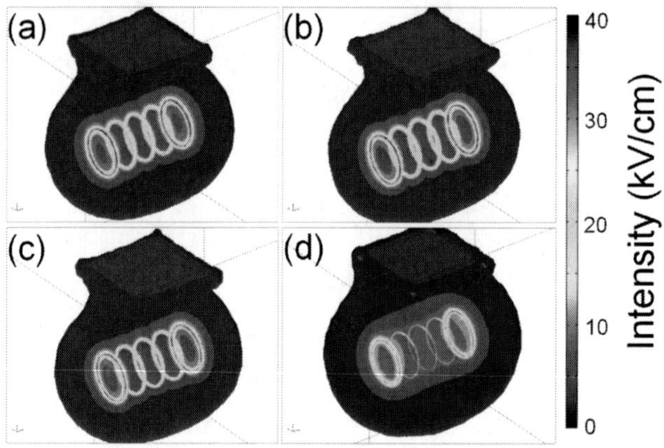

Figure 6.18 Electric field intensity profiles of multiple discs linked with a shaft with different shaft diameters: (a) 0 mm, (b) 20 mm, (c) 30 mm, (d) 38 mm (applied voltage = 60 kV, collecting distance = 13 cm).

Electric field intensities of multiple discs with different shaft diameters are shown in Fig. 6.19a. For cylinder (shaft diameter 40 mm), there is no intensity peak formed in the middle area. With the decrease in the shaft diameter, electric field peaks are generated on the disc top. With the shaft diameter decreasing from 40 to 30 mm, the electric field intensity increases evidently. Further decrease of the shaft diameter leads to slightly increased electric field intensity (Fig. 6.19a).

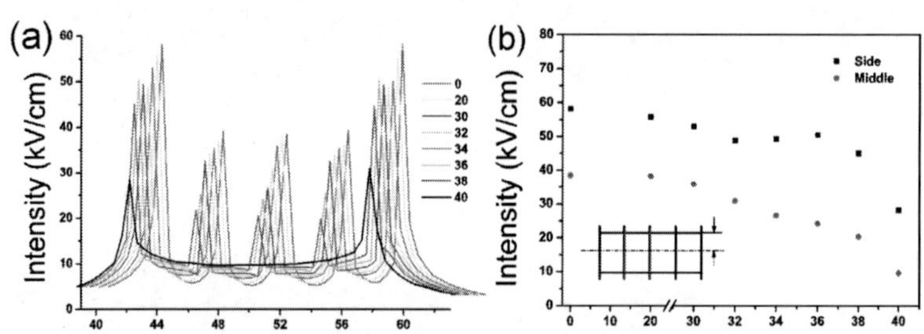

Figure 6.19 Electric field intensity of multiple discs linked with a shaft with different shaft diameters, (a) axial direction; (b) comparison of the intensity peaks (applied voltage = 60 kV, collecting distance = 13 cm).

Figure 6.19b shows the intensity results of both the middle and side discs with different shaft diameters. The intensity of the discs decreases gradually with the increase of the shaft diameter. The intensity drops to a very low level when the shaft diameter approaches 40 mm, under which the spinneret behaves more like a cylinder with uneven electric field distribution.

The inter-disc distance influences the electric field greatly. Figure 6.20 shows the electric field intensity profiles of spinnerets with different inter-disc distances. When the distance is too short, 0.5 cm for instance, no evident electric field intensity peaks are formed on the disc top (Fig. 6.20a). Many peaks can be observed when the distance is 1 cm, but the intensity of the peaks is very low. With the increase of the inter-disc distance, the number of peaks in the middle area decreases due to the disc number decrease; however, the intensity of the individual peak increases evidently.

Figure 6.20b shows the electric field intensity of the middle disc with different disc distance under different shaft diameter. It can be observed that with the increase of disc distance from 0 to 20 mm, the intensity increases evidently. Increasing the distance further from 20 to 40 mm leads to a slight increase of intensity. The smaller the shaft diameter, the higher the intensity and the higher the increase rate of the intensity. When the disc distance is too short, the whole spinneret behaves like a cylinder with uneven electric field. Increasing the inter-disc distance to 40 mm can greatly increase the electric field intensity of each disc.

As the disc distance decreases, more discs can be added into the liquid bath, and the productivity may increase. However, the electric field intensity must be high enough to insure polymer jets will be produced on each disc. Thus, a proper disc distance with higher electric field intensity and relatively higher productivity is important in this case.

6.5 Ring Electrospinning

The diameters of the ring and ring wire were 8 cm and 2 mm, respectively. The polymer bath was 24 cm in length, 10 cm in width, and 4 cm in height. A metal plate (24 cm long, 15 cm wide, and 5 mm thick) was used as a collector. The applied voltage and collecting distance were set at 60 kV and 13 cm, respectively.

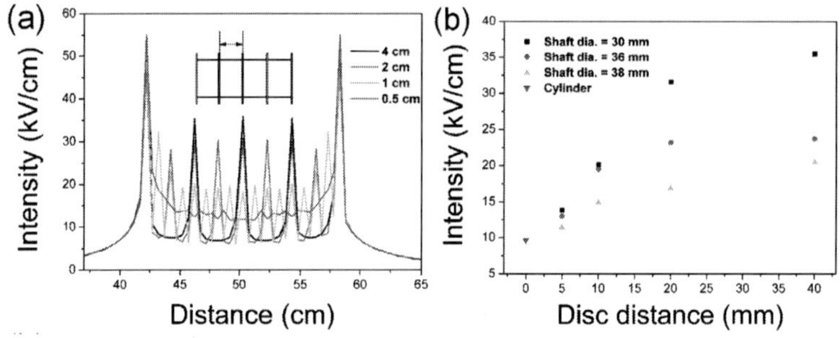

Figure 6.20 Electric field intensity of multiple discs with different shaft diameters, (a) axial direction (shaft diameter 30 mm); (b) comparison of the intensity peaks of the middle discs under different shaft diameters (applied voltage = 60 kV, collecting distance = 13 cm).

Figure 6.21 shows the electric field intensity profiles. High intensity electric field is just formed on the surface of the ring (Fig. 6.21a), and only the top half of the ring shows very high intensity electric field according to the magnified view in Fig. 6.21b. The strongest electric field appears on the very top of the ring. Because of the small radius of the ring wire, charges tend to be distributed on the narrow surface of the ring top. In a very wide area, the ring shows the electric field intensity higher than 50 kV/cm, as indicated by the red area in Fig. 6.21b. The bottom of the ring has similar electric field intensity to that of the polymer solution.

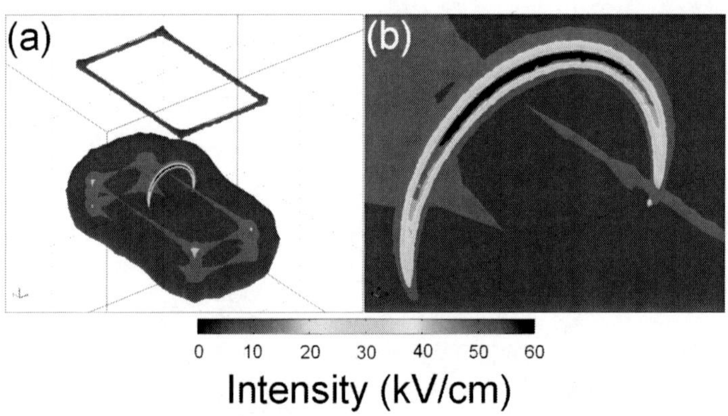

Figure 6.21 Electric field intensity profiles of a ring spinneret, (a) overview; (b) magnified view of the ring area (applied voltage = 60 kV, collecting distance = 13 cm).

It is obvious that ring coil can form concentrated electric field on the surface. When polymer solution covers the surface where high electric field is formed, solution jets will be generated due to the high electrostatic force the polymer solution is subject to. Since the surface of the ring shows very concentrated electric field, polymer jets will be produced from a very wide area of the ring, suggesting that the ring coil is an efficient needleless electrospinning spinneret.

The electric field intensity profile of the ring coil is shown in Fig. 6.22. The 3D view of the surface electric field intensity shows that high electric field intensity is formed around the ring surface with the highest intensity value appearing at 90° to the ring angle θ (Fig. 6.22a). From the axial direction, a very sharp peak is formed along the axial direction of the ring, and the intensity peak is higher than 70 kV/cm (Fig. 6.22b). Figure 6.22c shows the intensity at different ring angles. The intensity increases from the ring bottom to the top. Figure 6.22d shows the intensity from the ring center to the collector (along the spinning way). The high intensity was found mainly on the ring surface. This suggests that high electric field intensity is just formed around the ring coil surface.

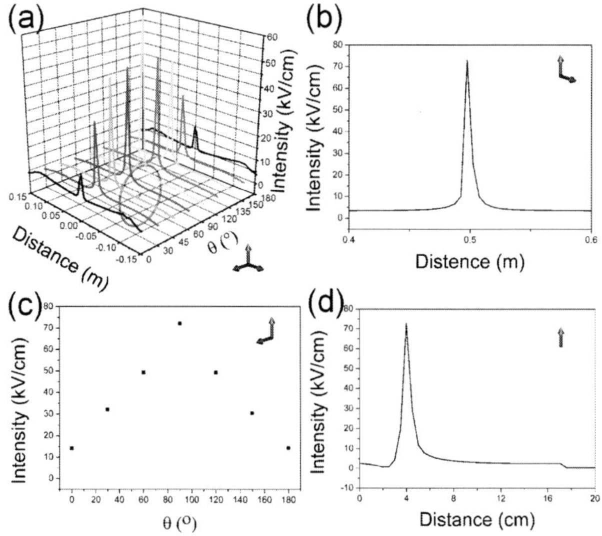

Figure 6.22 Electric field intensity of a ring spinneret, (a) 3D view; (b) axial direction of the ring; (c) along the surface of the ring; (d) along the spinning direction (applied voltage = 60 kV, collecting distance = 13 cm).

Figure 6.23 shows the electric field intensity profiles under different applied voltages. With an increase in the applied voltage, the electric field intensity along the top ring increases, as evidenced by the color change of the ring surface.

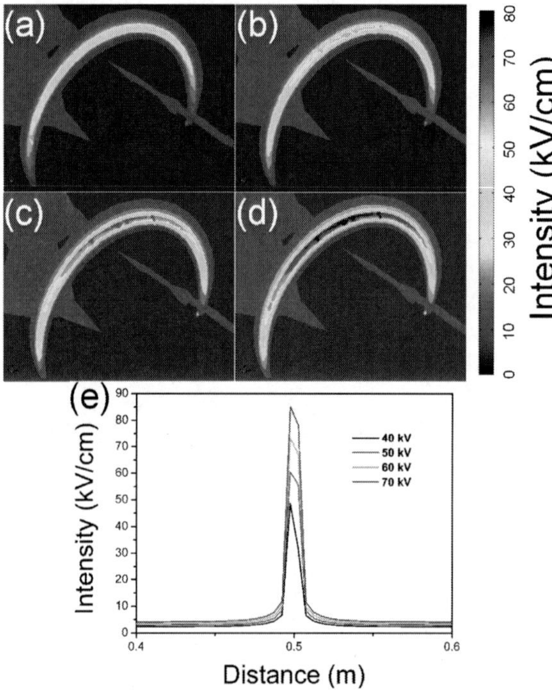

Figure 6.23 Electric field intensity profiles of ring spinneret under different applied voltages: (a) 40 kV, (b) 50 kV, (c) 60 kV, (d) 70 kV. (e) Electric field intensity comparison (collecting distance = 13 cm).

Figure 6.23e shows the intensity of the ring surface under different applied voltages. The peak value, which means the largest intensity, increases greatly with the increase of applied voltage. When applied voltage is 40 kV, high electric field forms on the top half of the ring with the electric field intensity around 40 kV/cm. With the increasing applied voltage, the top ring surface changes from dark yellow to light red, suggesting the increase in the electric field intensity to as high as 60 kV/cm. At 70 kV, a very wide red area with intensity over 70 kV/cm is formed on the ring surface. Therefore, increasing the applied voltage not only strengthens the electric field but also widens the area with high electric field intensity on the spinneret surface.

Figure 6.24 shows the electric field intensity of the ring coil with different collecting distances. When the collecting distance changes from 19 to 10 cm, the intensity increases from around 65 kV/cm to about 75 kV/cm.

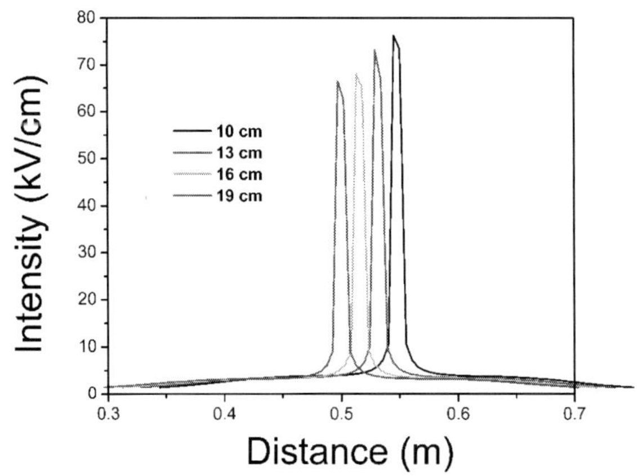

Figure 6.24 Electric field intensity of ring spinneret under different collecting distances (applied voltage = 60 kV).

It seems that increasing the collecting distance further may not affect the electric field much. However, extending a collecting distance is not applicable given that most fibers cannot be collected after the whipping process of electrospinning. Figure 6.25 shows the electric field intensity profiles of the ring spinneret with different wire diameters. With the increase of wire diameter, the electric field strength decreases gradually. As the electric field tends to concentrate on the surface of high curvature, the small diameter ring shows stronger electric field intensity under the same applied voltage.

In needleless electrospinning, increase of the nozzle area may enhance the fiber generation thus improving the productivity. A decrease of the electric field intensity leads to less polymer jets for the same area of the spinneret and the as-spun nanofibers may be coarser compared to that from spinneret with a highly concentrated electric field. It is a challenge to increase the spinnable area of the spinneret, which also has a highly concentrated electric field.

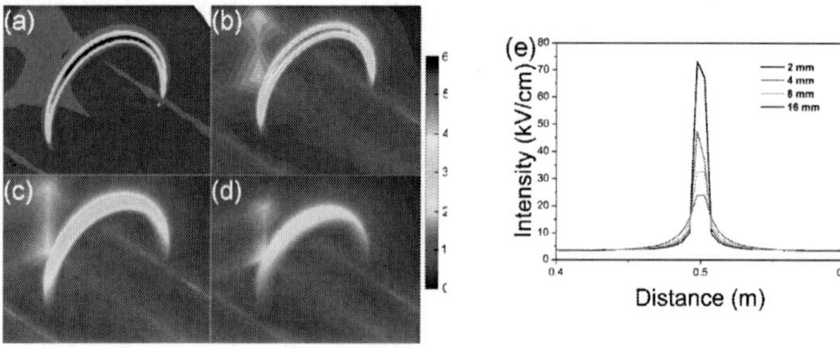

Figure 6.25 Electric field intensity of ring spinneret under different wire diameters: (a) 2 mm, (b) 4 mm, (c) 8 mm, and (d) 16 mm. (e) Electric field intensity comparison (applied voltage = 60 kV, collecting distance = 13 cm).

Figure 6.26 shows the electric field intensity profiles of the ring having different diameters. When the ring diameter increases, the ring is more isolated from the polymer solution, and the electric field intensity profile is very different compared with the polymer solution. As seen from Fig. 6.26, when the ring diameter is 2 cm, the electric field intensity is very weak compared to that of a ring with diameter of 8 cm.

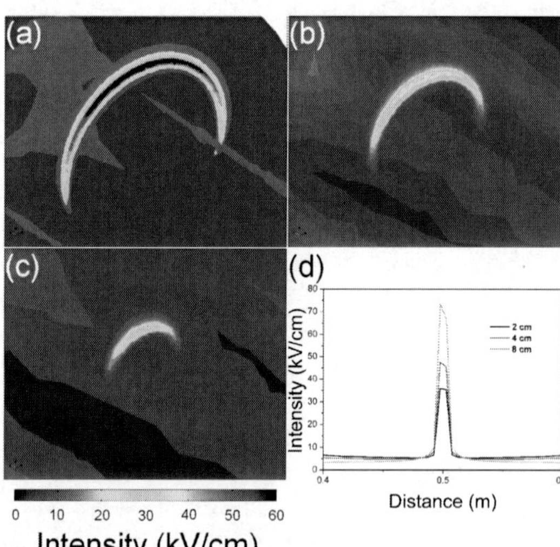

Figure 6.26 Electric field intensity of ring spinnerets under different ring diameters: (a) 8 cm, (b) 4 cm, and (c) 2 cm. (d) Electric field intensity comparison (applied voltage = 60 kV, collecting distance = 13 cm).

6.6 Multiple Rings

Electric field intensity profiles of the spinnerets with different numbers of rings are shown in Fig. 6.27.

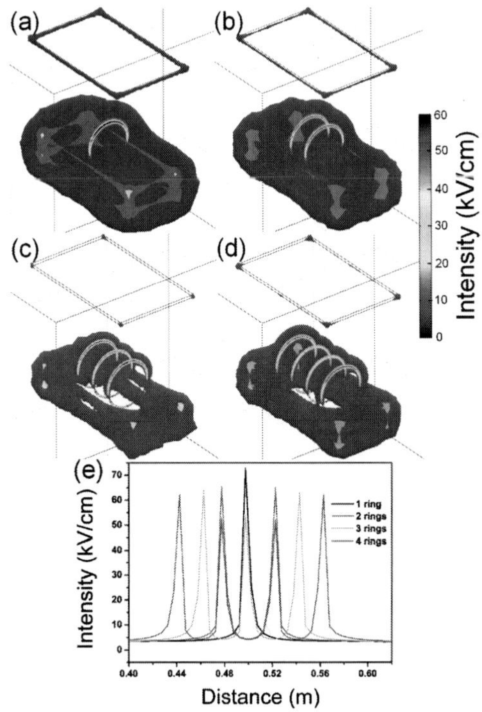

Figure 6.27 Electric field of rings with different ring number, (a–d) electric field intensity profiles and (e) electric field intensity from nozzle to collector (applied voltage = 60 kV, collecting distance = 13 cm, ring distance = 4 cm).

Strong electric field is formed on the top half of each ring with the intensity higher than 50 kV/cm. When the number of rings increases, it is evident that higher electric field is formed on both sides of the rings, but weaker on the middle ring surface. When a shaft is used to connect the multiple rings, the influences of shaft diameter and distance between rings show similar trends as those of the multiple discs described previously. Figure 6.27e shows the electric field intensity of the rings along the axial direction. It is shown that higher intensity electric field distributes on the side of the rings when the ring number increases and the intensity of middle rings decreases slightly.

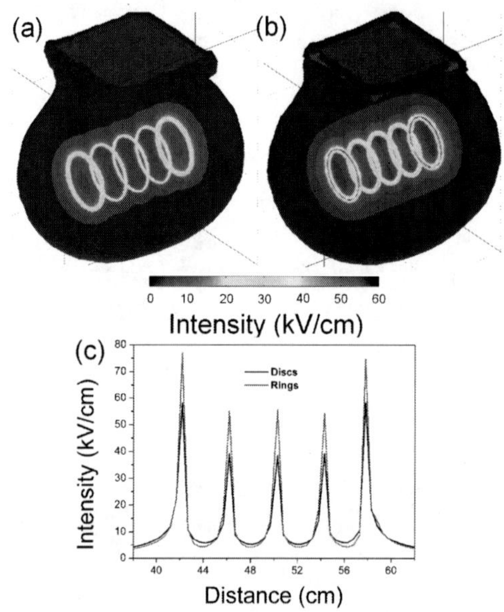

Figure 6.28 Electric field comparison of multiple discs and multiple rings, (a–b) electric field intensity profiles of multiple discs and multiple rings, respectively; (c) electric field intensity (applied voltage = 60 kV, collecting distance = 13 cm, disc/ring distance = 4 cm).

Figure 6.28 shows the comparison of electric field intensity of multiple discs and multiple rings under the same conditions. It is revealed that rings can generate stronger electric field compared to discs. As shown in Fig. 6.28c the intensity of multiple rings is much higher than that of multiple discs.

6.7 Conical Wire Coil Electrospinning

Multiple rings may lead to uneven fiber generation due to the "gaps" between the rings. A continuous ring structure, coil, is proposed to be a spinneret to avoid such a problem. Two coil structures, conical and spiral coils, are subjected to electric field intensity analysis.

The electric field intensity profiles of the conical coil and conventional needle are shown in Fig. 6.29. The conical coil forms concentrated electric field lines around the wire surface because of its small radius. The concentrated field lines are also formed between the adjacent wires, except that the field intensity is less than that on the wire surface. Similar to the conical coil spinneret, the needle nozzle also gives concentrated electric field lines at its tip.

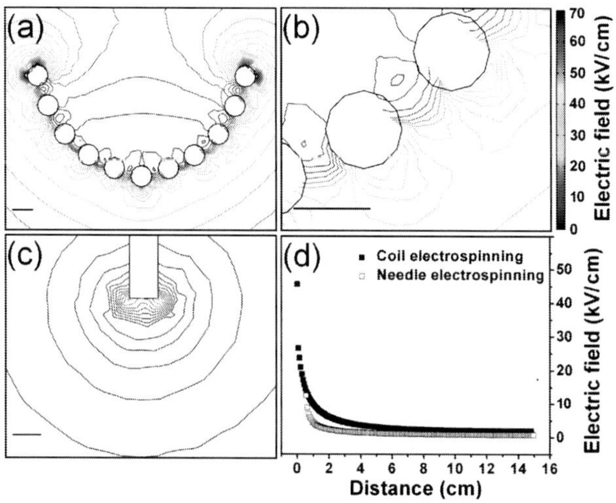

Figure 6.29 Cross-sectional view of electric field intensity profiles on (a, b) conical wire coil spinneret (applied voltage = 60 kV) and (c) needle nozzle (applied voltage = 22 kV); (d) electric field intensity along the electrospinning direction. Bar = 1 mm.

By comparing the color of the electric field lines, the conical exhibits higher intensity electric field than the needle. Figure 6.29d illustrates the electric field intensity along the spinning way and the conical wire coil shows much higher electric field intensity.

6.8 Tubular Wire Coil

Figure 6.30 illustrates the electric field intensity profiles of needle and tubular coil electrospinning systems. The tubular coil forms concentrated field lines around the coil surface because of its small radius. Compared to needle setup, each spiral of the coil shows a concentrated electric field around it. Figure 6.30 also shows the electric field intensity of both setups along the spinneret surface and along the spinning direction. Under the respective applied voltages, the electric field intensity of the coil setup is much higher than that of the needle setup along the spinning way, as shown in Fig. 6.30d. A front view (Fig. 6.30b) and electric field intensity analysis (Fig. 6.30c) show that high electric intensity peaks are generated on the top of each spiral, whereas the tip of the needle just shows one peak with lower intensity (Fig. 6.30c).

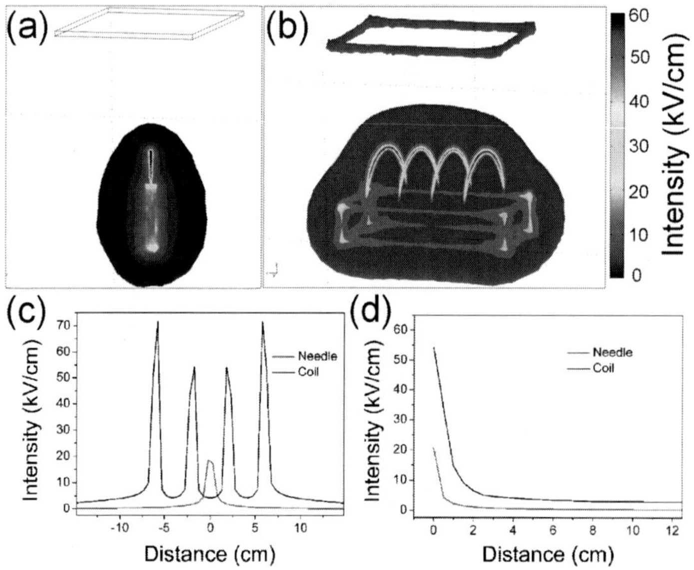

Figure 6.30 Electric field intensity profiles of (a) needle electrospinning and (b) tubular wire coil electrospinning. Electric field intensity of needle and tubular wire coil on axial direction (c) and spinning way (d) (applied voltage = 60 kV for coil and 22 kV for needle, collecting distance = 13 cm).

6.9 Comparison of Cylinder, Disc, Ring, and Coil

It is evident that needleless electrospinning spinnerets with different geometries form very different electric field/intensity when they were subjected to the same applied voltage. From cylinder to disc, ring and to continuous coil, the electric field intensity show a rising trend.

Figure 6.31 shows different spinnerets for needleless electrospinning. Table 5.1 lists the dimensions of those spinnerets, and the lengths of cylinder and tubular coil.

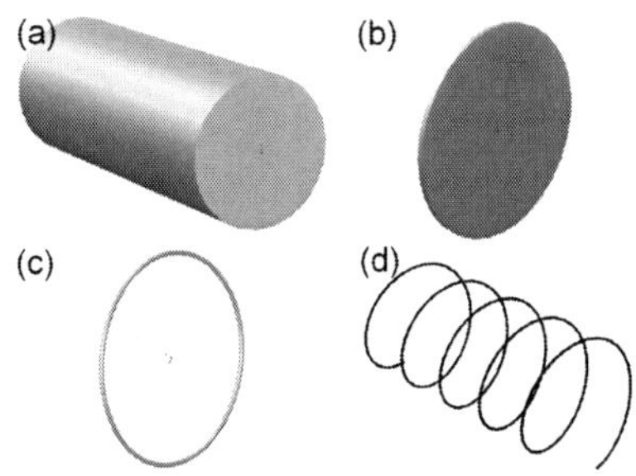

Figure 6.31 Schematics of different needleless electrospinning spinnerets: (a) cylinder, (b) disc, (c) ring, (d) tubular coil.

Table 6.2 Dimensions of different needleless electrospinning spinnerets

Nozzle	Dimension		
	Diameter (cm)	**Length (cm)**	**Wire diameter (mm)**
Cylinder	8	20	N/A
Disc		0.2	N/A
Ring		0.2	2
Coil		20	2

Figure 6.32 shows the electric field intensity profiles of different spinnerets under the same conditions. For cylinder spinneret, electric field shows uneven distribution on the surface. Higher electric field is formed on both ends of the cylinder while the electric field intensity on the middle area is lower.

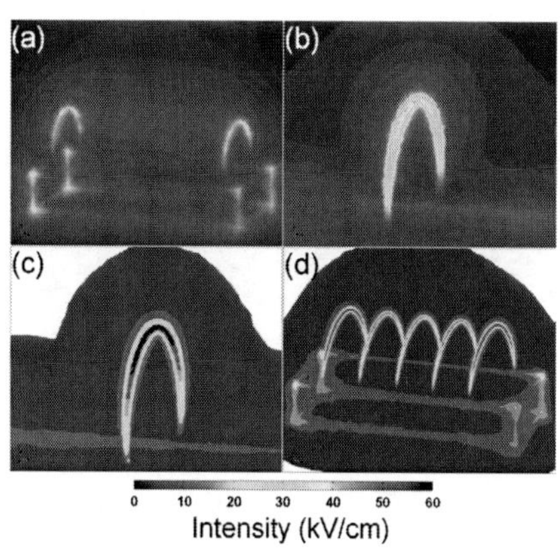

Figure 6.32 Electric field profiles of (a) cylinder, (b) disc, (c) ring, and (d) coil (applied voltage = 60 kV, collecting distance = 13 cm).

For the disc, much higher electric field is formed on the top circumference compared to that of cylinder ends. The ring shows the same electric field to the disc, except the intensity is much higher. Higher electric intensity is formed on the top of spiral coil. It is evident that ring and tubular coil show much higher intensity electric field than that of cylinder and disc spinnerets. Figure 6.33 shows the electric field intensity along the axial direction at different central angles. The highest electric field intensity is formed on the spinneret top. With the angle θ increases to 180° or decreases to 0°, the intensity decreases to almost zero.

For the cylinder, an uneven electric field is formed on the spinneret surface with higher intensity on both top ends, but no peaks found on the cylinder middle. Much higher electric field is formed on the disc edge, which suggests that disc shows higher efficiency to concentrate electric field. Further higher electric intensity peaks are formed on the ring top. The tubular coil shows the same electric intensity value with the ring, except more sets of peaks are formed.

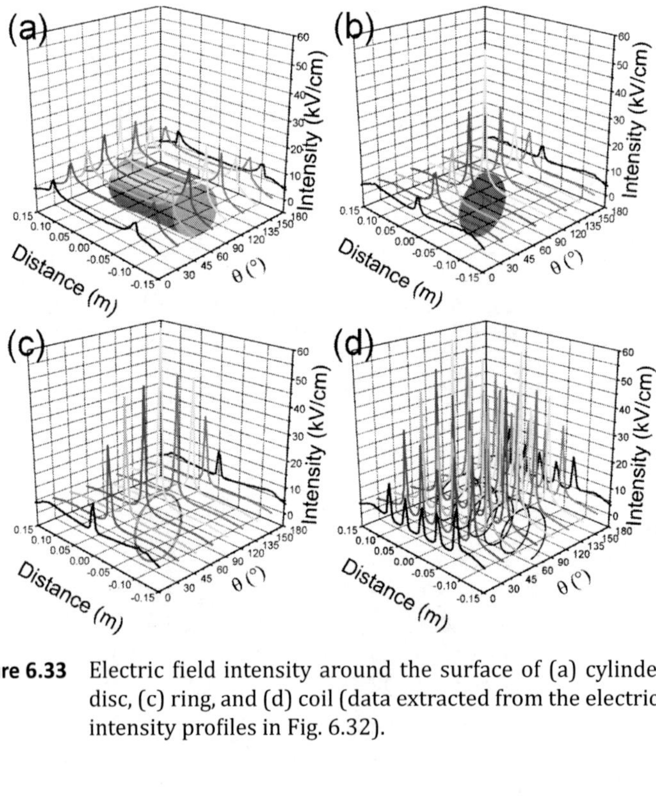

Figure 6.33 Electric field intensity around the surface of (a) cylinder, (b) disc, (c) ring, and (d) coil (data extracted from the electric field intensity profiles in Fig. 6.32).

Figure 6.34 illustrates the electric field intensity around the spinneret surface (from side view) and from the spinneret to collector. With the increase in the central angle, the cylinder middle keeps stable, which shows very low intensity. The intensity of the cylinder side is much higher than that of the cylinder middle, and the highest intensity appears at the angle of 90° which is the top of the cylinder. For the disc, the electric field intensity is higher. However, when wire coil is used, either ring or tubular coil, the intensity at 90° further increases to about 70 kV/cm.

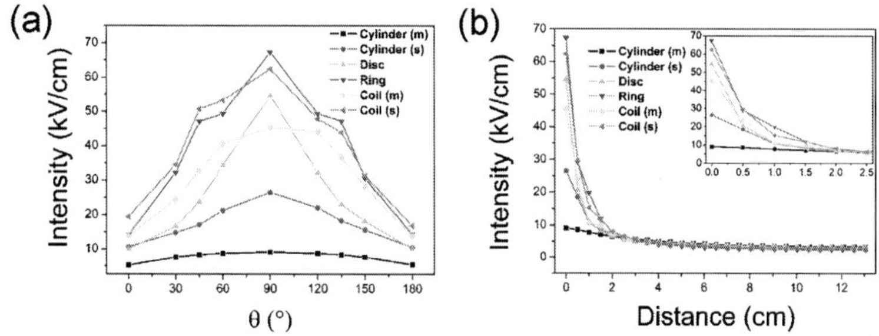

Figure 6.34 Electric field intensity of different spinnerets, (a) axial direction; (b) along the spinning direction. (Applied voltage = 60 kV; collecting distance = 13 cm).

The electric field intensity drops dramatically from the spinneret surface to the collector as shown in Fig. 6.34b. Wire coils show higher intensity compared to that of disc and cylinder, which suggests that a relatively high electric field is formed on the surface of the coil spinnerets. The electric field intensity of the spinneret surface determines the nanofiber production rate and the diameter and uniformity of the fiber produced; thus, higher production rate with more uniform and thin nanofibers can be produced by using wire coil spinnerets.

6.10 Collectors

When a stick is chosen as collector, the influence of collector length on the electric profile is studied. Increasing the collector length from 15 to 30 cm shows little impact on the electric field intensity.

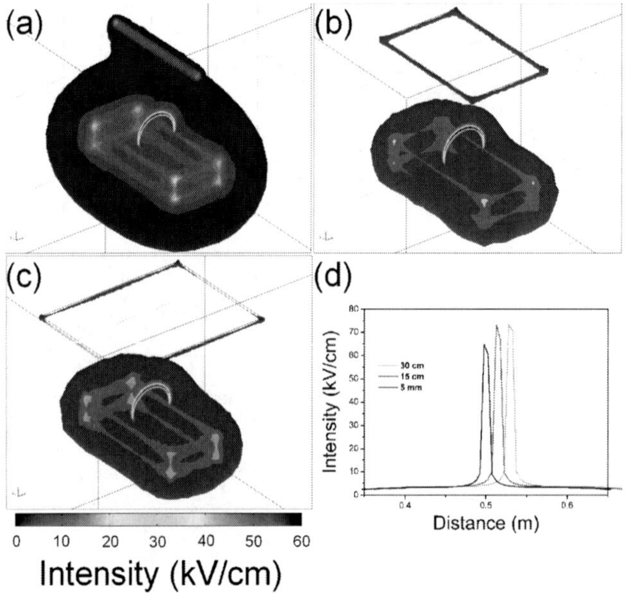

Figure 6.35 Electric field intensity of ring spinneret with different collector lengths (applied voltage = 60 kV, collecting distance = 13 cm).

It seems that change of collector shape has an influence on the electric field intensity. Fiber collection in electrospinning is further examined in the next chapter.

Chapter 7

Fiber Collection

The fiber collection in needleless electrospinning can generally be classified into three types: grounded plate, rotating drum, and moving substrate. The plate collector is normally used for electrospinning at a moderate production rate, e.g., multi-jet electrospinning and wire coil electrospinning. However, these fiber collectors are not suitable for upward needleless electrospinning systems, because the deposition of numerous nanofibers accumulates a large amount of charge on the collector, which can disturb or even stop the electrospinning process if they do not dissipate quickly. Moving collectors are therefore necessary in these systems. There are two types of moving collectors: rotating drum collector and moving substrate. The drum collector was used by Niu et al. [425]. Drum collector can also collect aligned nanofiber mat. When the mat reaches a certain thickness, it can be peeled off the collector. Jirsak et al. [435] used a moving substrate to collect nanofibers in their systems. The advantage of this type of fiber collection is that nanofibers can be collected continuously and the nanofiber web does not have to be very strong; such a collection system also allows the incorporation of nanofibers into the fiber collecting substrate itself, if required.

7.1 Effect of Collector Shape

Different collectors, including stick, plate, arc, and rotating drum were used to collect nanofibers. Fiber diameter is shown in Fig. 7.1a. When

Needleless Electrospinning of Nanofibers: Technology and Applications
Tong Lin and Xungai Wang

ISBN 978-981-4316-35-4 (Hardcover), 978-981-4316-84-2 (eBook)
www.panstanford.com

a stick collector was used, the electrospun nanofibers had an average diameter of 300 nm, while nanofibers collected by other collectors were finer, around 250 nm in average diameter. Productivity results are shown in Fig. 7.1b. The productivity was very low when using a stick as a collector. The collecting area increased for drum, plate, or arc, and the productivity increased accordingly, to over 12 g/h. Increasing the collecting area led to a larger spinneret surface to generate nanofibers, thus increasing the nanofiber production rate.

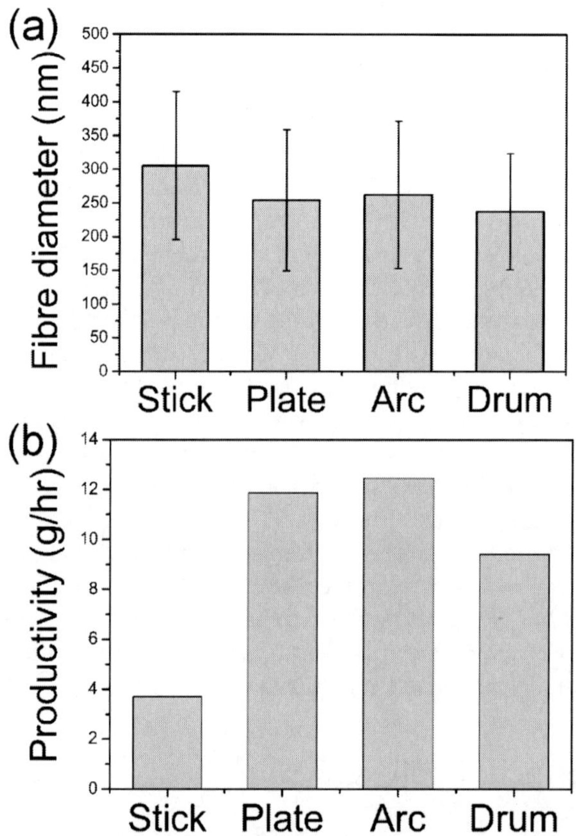

Figure 7.1 Effect of collector shape on needleless electrospinning, (a) fiber diameter; (b) productivity. (PVA concentration = 9%, applied voltage = 60 kV, collecting distance = 13 cm).

It was noted that the nanofibers accumulated on the collector to form a transparent film. For stick, plate, or arc, the collected fibers were observed to merge together due to the high content of solvent remaining in the fibers. When a rotating drum was used, semi-dry nanofibers were collected.

7.2 Effect of Collection Media

In order to accelerate the evaporation of solvent in needleless electrospinning, different collection media, including aluminum foil, fabric and meshes with different pore sizes, were used to cover the drum surface. Fiber diameter was tested as shown in Fig. 7.2. The average fiber diameter was around 250 nm, and changed little for the different collectors used. When an aluminum foil was used to collect nanofibers, the fibers interconnected with each other and fibers stuck together forming blocks. The fiber mat collected by the aluminum foil was brittle and fragile, and behaved like a transparent membrane. When a fabric was used to collect nanofibers, the overall fibrous structure was improved because of the good air permeability of the fabric, which enabled the solvent in the wet fibers to be easily evaporated.

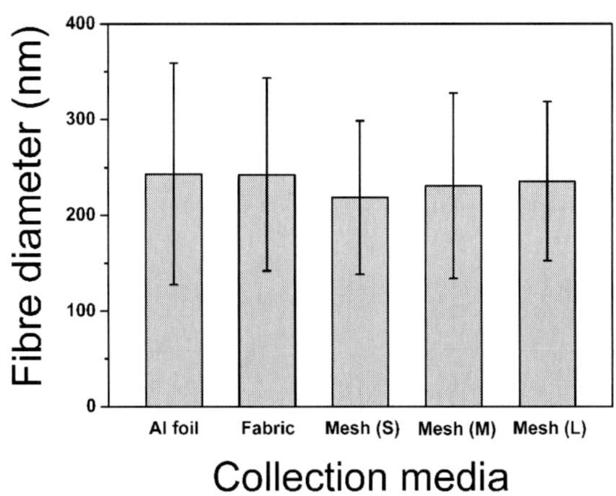

Figure 7.2 Effect of collection media on the fiber diameter of needleless electrospun nanofibers. (PVA concentration = 9%, applied voltage = 60 kV, collecting distance = 13 cm).

The nanofiber mat collected by fabric was similar to that collected by metal sheet. Also, the resultant product was flexible and felt like a nonwoven mat. Mesh with different pore sizes can be used to collect the electrospun nanofibers. When the pore size was very small, an improved fibrous structure was obtained. However, with the increase in pore size, the collected nanofiber mat became uneven. When the pore size was 1 cm^2, fiber patterns were formed.

7.3 Effect of Collecting Distance

Figure 7.3 shows the morphology of the electrospun nanofibers (PVA concentration = 9 wt%) collected at different collecting distances. The majority of fibers were on nano-scale, and they were collected in the form of a nonwoven structure. It was observed that a shorter collecting distance led to finer nanofibers because of the more concentrated electric field. However, a shorter collecting distance resulted in shorter time for the solvent to evaporate from the filaments. As a result, semi-dry nanofibers were collected, which made the fibers stick together forming a bridged and interconnected fibrous structure (Fig. 7.3a). By contrast, nanofibers collected over a longer distance showed isolated fiber morphology (Fig. 7.3c). Due to the decreased electric field intensity in this situation, the as-spun fibers were coarser.

The fiber diameter increased and productivity decreased with the increase in the collecting distance. The main reason for this is the decreased electric field intensity. A moderate collecting distance is suggested to prevent "corona discharge" during electrospinning, and to obtain a fibrous structure at a relatively high production rate.

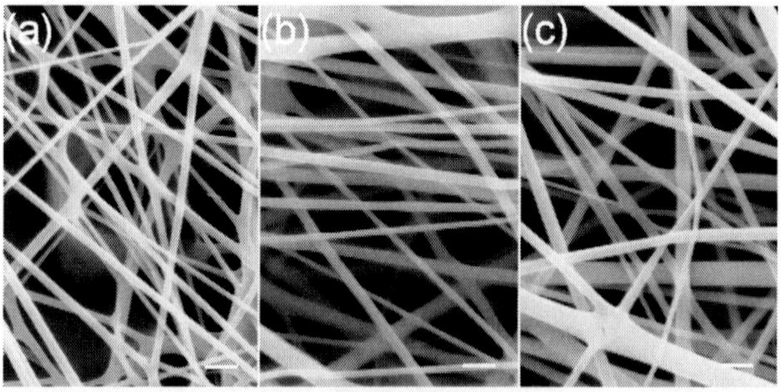

Figure 7.3 SEM images of needleless electrospun nanofibers with different collecting distances, (a) 13 cm, (b) 15.5 cm, and (c) 18 cm. Bar = 1 μm. (PVA concentration = 9%, applied voltage = 60 kV).

7.4 Effect of Air Flow

Air flow has been used to modify needle electrospinning, and the as-spun nanofibers showed reduced diameter. For needleless electrospinning, using air flow may accelerate the solvent evaporation and thus reduce the fiber–fiber interconnections. Air flow was first inserted into the center of the ring coil to modify the electrospinning process. Figure 7.4 shows the SEM images of the as-spun nanofibers at different magnifications (×2000 and ×10000).

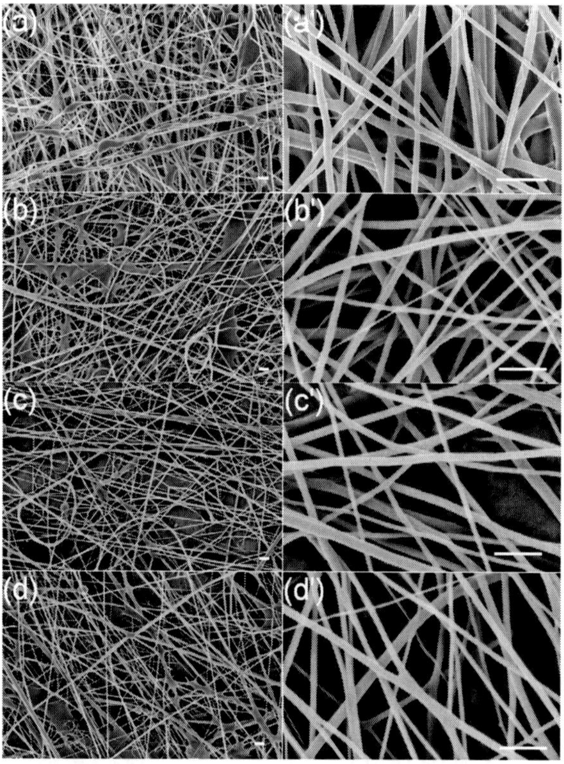

Figure 7.4 SEM images of nanofibers electrospun with different air flow (center of the ring) pressures, (a) and (a′) control; (b) and (b′) 10 psi; (c) and (c′) 15 psi; (d) and (d′) 20 psi. Bar = 2 μm. (PVA concentration = 9%, applied voltage = 60 kV, collecting distance = 15 cm).

Without air flow, nanofibers tended to stick to each other (Fig. 7.4a). However, when air flow was applied to the center of the coil, fewer blocks were observed within the nanofiber mats (Fig. 7.4b–d). Figure 7.4b′–d′ shows a clear fibrous structure and isolated nanofibers. These results clearly indicated that air flow in the center of the ring accelerated the solvent evaporation, which prevented the fiber–fiber interconnections. Figure 7.5a shows the fiber diameter and its distribution. The inset shows the diameter distribution when air flow with a pressure of 10 psi was applied. A narrower diameter distribution, with a mean diameter around 200 nm, was achieved when air flow was applied. With the increase in the air flow pressure, the mean diameter of the as-spun nanofibers decreased slightly and was much smaller than that of the control. Also the diameter distribution was much narrower than that of the control. When air flow was applied in the center of the ring coil, the air flow might assist in stretching the polymer jets thus producing finer nanofibers.

Figure 7.5b shows the productivity of nanofibers for the air-flow-enhanced electrospinning. The productivity decreased when air flow was applied. The reason for this is that when the air flow blew the ring where polymer jets were produced, the surface of the coated solution on the top ring changed. Some of the fiber-generating areas were affected and thus affecting the jet formation. It was noted that when the pressure of air flow was 20 psi, most polymer solution was thrown out because of the strong air flow. Polymer solution blocks could be observed with the naked eye on the collector and thus the productivity was higher than the control.

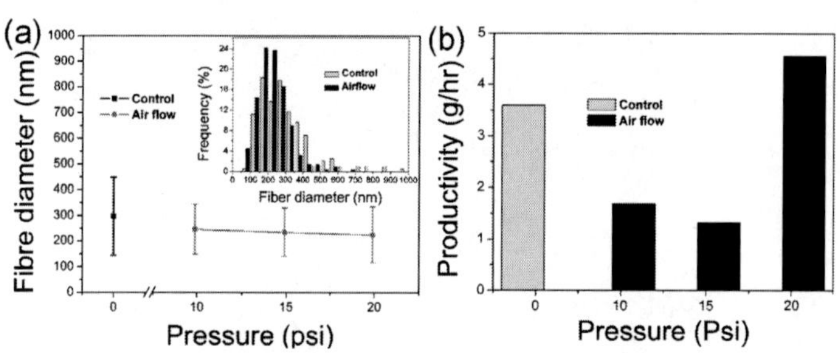

Figure 7.5 Effect of air flow (center of the ring) pressure on the fiber diameter (a), diameter distribution (10 psi) (inset of (a)), and productivity (b) of as-spun nanofibers. (PVA concentration = 9%, applied voltage = 60 kV, collecting distance = 15 cm).

Figure 7.6 shows the SEM images of nanofibers electrospun from needleless electrospinning with air flow along the spinning direction. Compared to Fig. 7.6a, these nanofibers showed better fibrous morphology with less fiber–fiber interconnection and fewer fiber blocks. The nanofibers produced were quite even and most of the fibers were isolated from each other.

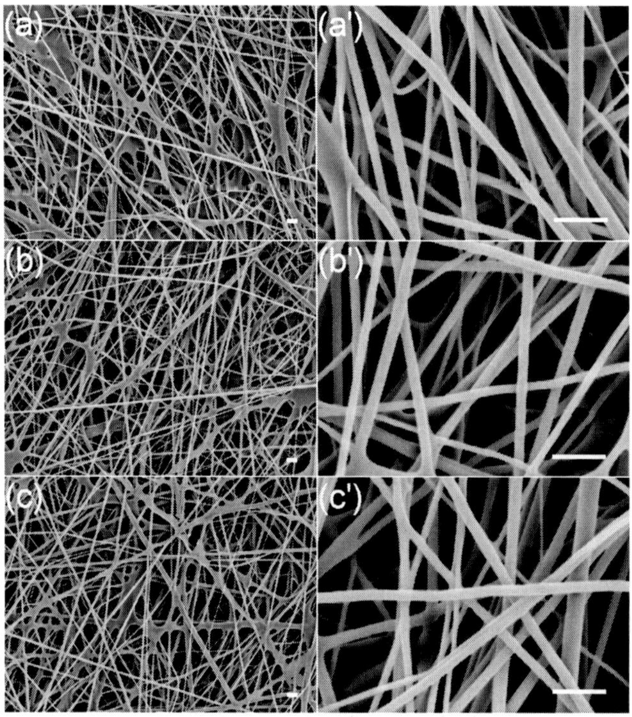

Figure 7.6 SEM images of nanofibers electrospun with different air flow (along the spinning direction) pressures, (a) and (a′) 10 psi; (b) and (b′) 15 psi; (c) and (c′) 20 psi. Bar = 2 μm. (PVA concentration = 9%, applied voltage = 60 kV, collecting distance = 15 cm).

It is obvious that applying air flow along the spinning direction can reduce the fiber–fiber interconnections. The air flow along the spinning direction accelerated the solvent evaporation and thus improved the fiber uniformity, which was the same as air flow from the center of the ring coil. However, increasing the air flow pressure had less impact on the nanofiber morphology as seen in the SEM images. Air flow along the spinning direction did not affect the fiber-stretching process.

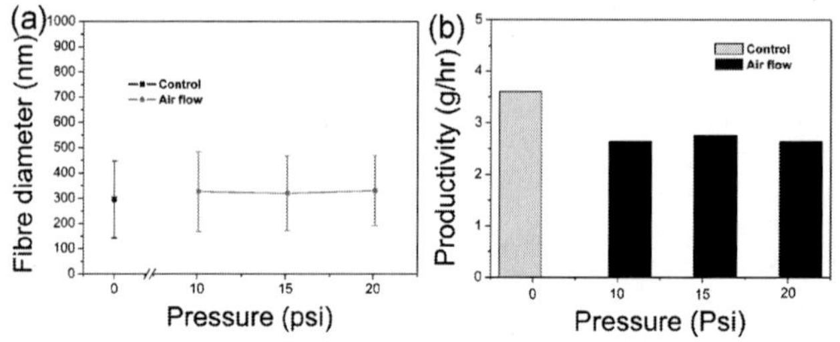

Figure 7.7 Effect of air flow (along the spinning direction) pressure on the fiber diameter (a) and productivity (b) of as-spun nanofibers. (PVA concentration = 9%, applied voltage = 60 kV, collecting distance = 15 cm).

Fiber diameter results of the as-spun nanofibers are shown in Fig. 7.7a. It is evident that applying the air flow along the spinning direction had no impact on the fiber diameter. Productivity of needleless electrospinning with air flow along the spinning direction is shown in Fig. 7.7b. The productivity decreased when air flow was used. This is probably due to insufficient fiber collection under the air flow.

Air flow was also applied at the collector with the SEM images of the collected nanofibers shown in Fig. 7.8. The fiber diameter and fiber productivity are shown in Table 7.1. The quality of the collected nanofibers is improved as seen from the SEM images. This is due to the accelerated solvent evaporation.

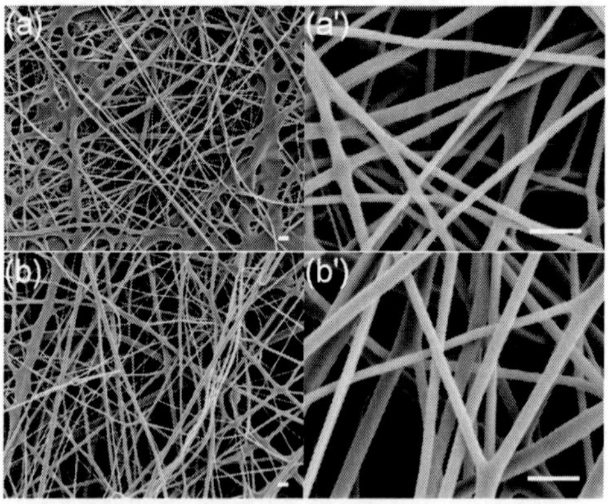

Figure 7.8 SEM images of nanofibers with ((b) and (b′)) and without ((a) and (a′)) air flow. Bar = 2 μm. (PVA concentration = 9%, applied voltage = 75 kV, collecting distance = 18 cm).

Since the air flow at the collector had no effect on the electric field, the fiber diameter and diameter distribution were almost the same as that of the control. The productivity also remained unchanged when air flow was employed at the collector. Since a longer collecting distance (18 cm) was chosen in this case, the productivity was much less than other cases discussed in this section.

Table 7.1 Fiber diameter and productivity of needleless electrospinning before and after air flow was applied to the collector (applied voltage = 75 kV, collecting distance = 18 cm)

Items	Control	Air flow
Fiber diameter (nm)	376.6 ± 179.1	406.4 ± 202.3
Productivity (g/h)	2.28	2.04

Chapter 8

Conclusion and Outlook

The former chapters of this book have shown that electrospun nanofibers have diverse applications in both textile and non-textile areas. Rotating coil may form a highly efficient needleless electrospinning system for mass production of nanofibers. In comparison to other rotating fiber generators, coil spinneret has much stronger and more concentrated electric field intensity in the fiber generating area. This is the main reason as to why coil electrospinning has higher fiber productivity with a much better control toward the fiber quality when compared with existing needleless electrospinning systems. A coil system with multiple rows of long coil spinnerets can produce continuous nanofiber mats of sufficient thickness and width in a single process. A coil electrospinning system can be used as a standalone machine for nanofiber production or incorporated into an existing production line to produce nanofiber-functionalized nonwovens, filter papers, and other fibrous products.

Our studies have shown that the finite element method can be used to understand the electrical field profiles in electrospinning, which are difficult to measure directly in a high electric field environment. By combining the finite element analysis with experimental results, the coil electrospinning can be further optimized and new electrospinning systems may be designed also.

Much work on electrospinning has been focused on the fiber generators, rather than the fiber collectors. The fiber collectors play

Needleless Electrospinning of Nanofibers: Technology and Applications
Tong Lin and Xungai Wang

ISBN 978-981-4316-35-4 (Hardcover), 978-981-4316-84-2 (eBook)
www.panstanford.com

a critical role in determining the final nanofibrous structures. More works need to be done on fiber deposition to effectively control the fiber orientation in the fibrous structure. It is also a challenge to electrospin nanofibers into yarns, an important building block to construct intricate fibrous structures other than the simple nonwovens. This would also require accurate control of the number of nanofibers from a needleless electrospinning system. At present, the exact number of fibers generated in a needleless electrospinning process is unknown.

In summary, needleless electrospinning represents an important technology for mass production of nanofibers. This technology will greatly facilitate the applications of nanofibers in different areas. Further work is still required to precisely control the fiber properties and the nanofibrous structures in needleless electrospinning. In addition, new needleless electrospinning systems and novel fiber collectors should be designed to produce nanofiber yarns, which will further expand the applications of nanofiber materials.

References

1. Supaphol P, Mit-uppatham C, Nithitanakul M. Ultrafine electrospun polyamide-6 fibers: effects of solvent system and emitting electrode polarity on morphology and average fiber diameter. *Macromolecular Materials and Engineering.* 2005; 290(9): 933–942.
2. Alborzi S, Lim L-T, Kakuda Y. Electrospinning of sodium alginate-pectin ultrafine fibers. *Journal of Food Science.* 2010; 75(1): C100–C700.
3. Bellan LM, Coates GW, Craighead HG. Poly(dicyclopentadiene) submicron fibers produced by electrospinning. *Macromolecular Rapid Communications.* 2006; 27(7): 511–515.
4. Keun Kwon I, Kidoaki S, Matsuda T. Electrospun nano- to microfiber fabrics made of biodegradable copolyesters: structural characteristics, mechanical properties and cell adhesion potential. *Biomaterials.* 2005; 26(18): 3929–3939.
5. Givens SR, Gardner KH, Rabolt JF, Chase DB. High-temperature electrospinning of polyethylene microfibers from solution. *Macromolecules.* 2007; 40(3): 608–610.
6. Ma PX, Zhang R. Synthetic nano-scale fibrous extracellular matrix. *Journal of Biomedical Materials Research.* 1999; 46(1): 60–72.
7. Liu G, Qiao L, Guo A. Diblock copolymer nanofibers. *Macromolecules.* 1996; 29(16): 5508–5510.
8. Hartgerink JD, Beniash E, Stupp SI. Self-assembly and mineralization of peptide-amphiphile nanofibers. *Science* (Washington, DC, United States). 2001; 294(5547): 1684–1688.
9. Martin CR. Membrane-Based Synthesis of Nanomaterials. *Chemistry of Materials.* 1996; 8(8): 1739–1746.
10. Ward GF. Meltblown nanofibres for nonwoven filtration applications. *Filtration and Separation.* 2001; 38(9): 42–43.

11. Weinberg M, Dee G, Harding T. Inventors, Flash spun web containing sub-micron filaments and process for forming same. *US Patent* 20060135020. 2006.
12. Fedorova N, Pourdeyhimi B. High strength nylon micro- and nanofiber based nonwovens via spunbonding. *Journal of Applied Polymer Science.* 2007; 104(5): 3434–3442.
13. Doshi J, Reneker DH. Electrospinning process and applications of electrospun fibers. *Journal of Electrostatics.* 1995; 35(2–3): 151–160.
14. Srinivasan G, Reneker DH. Structure and morphology of small diameter electrospun aramid fibers. *Polymer International.* 1995; 36(2): 195–201.
15. William G. *De Magnete* (Translated by P. E. Mottelay); 1600.
16. Rayleigh L. The influence of electricity on colliding water drops. *Proceedings of the Royal Society of London.* 1879; 28: 406–409.
17. Rayleigh L. On the capillary phenomena of jets. *Proceedings of the Royal Society of London.* 1879; 29: 71–79.
18. Zeleny J. The electrical discharge from liquid points, and a hydrostatic method of measuring the electric intensity at their surfaces. *Physical Review.* 1914; 3(2): 69–91.
19. Formhals A. Inventor process and apparatus for preparing artificial threads. *US Patent* 1,975,504. 1934.
20. Formhals A. Inventor production of artifical fibers from fiber forming liquids. *US Patent* 2,323,025. 1943.
21. Formhals A. Inventor method and apparatus for spinning. *US Patent* 2,349,950. 1944.
22. Taylor G. Disintegration of water drops in an electric field proceedings of the Royal Society of London Series A. *Mathematical and Physical Sciences.* 1964; 280: 383–397.
23. Taylor G. Electrically driven jets proceedings of the Royal Society of London Series A. *Mathematical and Physical Sciences.* 1969; 313: 453–475.
24. Kowalewski TA, Hiller WJ, Behnia M. An experimental study of evaporating small diameter jets. *Physics of Fluids A: Fluid Dynamics.* 1993; 5(8): 1883–1890.
25. Subbiah T, Bhat GS, Tock RW, Parameswaran S, Ramkumar SS. Electrospinning of nanofibers. *Journal of Applied Polymer Science.* 2005; 96(2): 557–569.

26. Reneker DH, Yarin AL, Fong H, Koombhongse S. Bending instability of electrically charged liquid jets of polymer solutions in electrospinning. *Journal of Applied Physics.* 2000; 87(9, Pt. 1): 4531–4547.
27. Dzenis Y. Material science: spinning continuous fibers for nanotechnology. *Science* (Washington, DC, United States). 2004; 304(5679): 1917–1918.
28. Yarin AL, Koombhongse S, Reneker DH. Bending instability in electrospinning of nanofibers. *Journal of Applied Physics.* 2001; 89(5): 3018–3026.
29. Hohman MM, Shin M, Rutledge G, Brenner MP. Electrospinning and electrically forced jets. I. Stability theory. *Physics of Fluids.* 2001; 13(8): 2201–2220.
30. Hohman MM, Shin M, Rutledge G, Brenner MP. Electrospinning and electrically forced jets. II. Applications. *Physics of Fluids.* 2001; 13(8): 2221–2236.
31. Yarin AL, Koombhongse S, Reneker DH. Taylor cone and jetting from liquid droplets in electrospinning of nanofibers. *Journal of Applied Physics.* 2001; 90(9): 4836–4846.
32. Lin T, Wang H, Wang H, Wang X. The charge effect of cationic surfactants on the elimination of fibre beads in the electrospinning of polystyrene. *Nanotechnology.* 2004; 15(9): 1375–1381.
33. Lin T, Wang H, Wang X. Self-crimping bicomponent nanofibers electrospun from polyacrylonitrile and elastomeric polyurethane. *Advanced Materials.* 2005; 17(22): 2699–2703.
34. Sun Z, Zussman E, Yarin AL, Wendorff JH, Greiner A. Compound core-shell polymer nanofibers by co-electrospinning. *Advanced Materials.* 2003; 15(22): 1929–1932.
35. Shin YM, Hohman MM, Brenner MP, Rutledge GC,. Experimental characterization of electrospinning: the electrically forced jet and instabilities. *Polymer.* 2001; 42(25): 09955–09967.
36. Liu H, Hsieh YL. Ultrafine fibrous cellulose membranes from electrospinning of cellulose acetate. *Journal of Polymer Science Part B: Polymer Physics.* 2002; 40(18): 2119–2129.
37. Koski A, Yim K, Shivkumar S. Effect of molecular weight on fibrous PVA produced by electrospinning. *Materials Letters.* 2004; 58(3–4): 493–497.
38. Gupta P, Trenor SR, Long TE, Wilkes GL. In situ photo-cross-linking of cinnamate functionalized poly(methyl methacrylate-co-2-hydroxyethyl

acrylate) fibers during electrospinning. *Macromolecules.* 2004; 37(24): 9211–9218.

39. Moon S, Choi J, Farris RJ. Preparation of aligned polyetherimide fiber by electrospinning. *Journal of Applied Polymer Science.* 2008; 109(2): 691–694.
40. Theron SA, Zussman E, Yarin AL. Experimental investigation of the governing parameters in the electrospinning of polymer solutions. *Polymer.* 2004; 45(6): 2017–2030.
41. Veleirinho B, Rei MF, Lopes-DA-Silva JA. Solvent and concentration effects on the properties of electrospun poly(ethylene terephthalate) nanofiber mats. *Journal of Polymer Science Part B: Polymer Physics.* 2008; 46(5): 460–471.
42. Lee K, Lee B, Kim C, Kim H, Kim K, Nah C. Stress-strain behavior of the electrospun thermoplastic polyurethane elastomer fiber mats. *Macromolecular Research Field.* 2005; 13(5): 441–445.
43. Inai R, Kotaki M, Ramakrishna S. Structure and properties of electrospun PLLA single nanofibres. *Nanotechnology.* 2005; 16(2): 208–213.
44. Jamila S, Changmo S. Electrospinning of polycarbonate nanofibers with solvent mixtures THF and DMF. *Journal of Materials Science.* 2004; 39(14): 4605–4613.
45. Bognitzki M, Czado W, Frese T, Schaper A, Hellwig M, Steinhart M, et al. Nanostructured fibers via electrospinning. *Advanced Materials.* 2001; 13(1): 70–72.
46. Casper CL, Stephens JS, Tassi NG, Chase DB, Rabolt JF. Controlling surface morphology of electrospun polystyrene fibers: effect of humidity and molecular weight in the electrospinning process. *Macromolecules.* 2003; 37(2): 573–578.
47. Kim C, Park S, Lee W, Yang K. Characteristics of supercapaitor electrodes of PBI-based carbon nanofiber web prepared by electrospinning. *Electrochimica Acta.* 2004; 50(2–3): 877–881.
48. Fong H, Liu W, Wang C-S, Vaia RA. Generation of electrospun fibers of nylon 6 and nylon 6-montmorillonite nanocomposite. *Polymer.* 2002; 43(3): 775–780.
49. Lee KH, Kim HY, La YM, Lee DR, Sung NH. Influence of a mixing solvent with tetrahydrofuran and N,N-dimethylformamide on electrospun poly(vinyl chloride) nonwoven mats. *Journal of Polymer Science Part B: Polymer Physics.* 2002; 40(19): 2259–2268.
50. Fong H, Reneker DH. Elastomeric nanofibers of styrene-butadiene-styrene triblock copolymer. *Journal of Polymer Science Part B: Polymer Physics.* 1999; 37(24): 3488–3493.

51. Deitzel JM, Kosik W, McKnight SH, Beck Tan NC, DeSimone JM, Crette S. Electrospinning of polymer nanofibers with specific surface chemistry. *Polymer.* 2002; 43(3): 1025–1029.

52. Matthews JA, Wnek GE, Simpson DG, Bowlin GL. Electrospinning of collagen nanofibers. *Biomacromolecules.* 2002; 3(2): 232–238.

53. McManus MC, Boland ED, Koo HP, Barnes CP, Pawlowski KJ, Wnek GE, et al. Mechanical properties of electrospun fibrinogen structures. *Acta Biomaterialia.* 2006; 2(1): 19–28.

54. Min B-M, Lee G, Kim SH, Nam YS, Lee TS, Park WH. Electrospinning of silk fibroin nanofibers and its effect on the adhesion and spreading of normal human keratinocytes and fibroblasts in vitro. *Biomaterials.* 25(7–8): 1289–1297.

55. Noh HK, Lee SW, Kim J-M, Oh J-E, Kim K-H, Chung C-P, et al. Electrospinning of chitin nanofibers: degradation behavior and cellular response to normal human keratinocytes and fibroblasts. *Biomaterials.* 2006; 27(21): 3934–3944.

56. Ohkawa K, Cha D, Kim H, Nishida A, Yamamoto H. Electrospinning of chitosan. *Macromolecular Rapid Communications.* 2004; 25(18): 1600–1605.

57. Muhammad Miftahul M, Hendri W, Ferry I, Kikuo O. Patterned indium tin oxide nanofiber films and their electrical and optical performance. *Nanotechnology.* 2008; (37): 375601.

58. Boland E, Wnek G, Simpson D, Pawlowski K, Bowlin G. Tailoring tissue engineering scaffolds using electrostatic processing techniques: a study of poly (glycolic acid) electrospinning. *Journal of Macromolecular Science, Part A.* 2001; 38(12): 1231–1243.

59. Hsu C. Preparation of biodegradable polyhydroxybutyrate fibers via electrospinning. 2007.

60. Caracciolo P, Thomas V, Vohra Y, Buffa F, Abraham G. Electrospinning of novel biodegradable poly(ester urethane)s and poly(ester urethane urea)s for soft tissue-engineering applications. *Journal of Materials Science: Materials in Medicine.* 2009; 20(10): 2129–2137.

61. Shu-Ying Gu, Qi-Lin Wu, Jie Ren, and Vancso, G. Julius. Mechanical properties of a single electrospun fiber and its structures. *Macromolecular Rapid Communications.* 2005; 26(9): 716–720.

62. Chen D, Liu T, Zhou X, Tjiu W, Hou H. Electrospinning fabrication of high strength and toughness polyimide nanofiber membranes containing multiwalled carbon nanotubes. *The Journal of Physical Chemistry B.* 2009, 113 (29): 9741–9748.

63. Ma Z, Kotaki M, Yong T, He W, Ramakrishna S. Surface engineering of electrospun polyethylene terephthalate (PET) nanofibers towards development of a new material for blood vessel engineering. *Biomaterials.* 2005; 26(15): 2527–2536.

64. Khil M, Kim H, Kim M, Park S, Lee D. Nanofibrous mats of poly (trimethylene terephthalate) via electrospinning. *Polymer.* 2004; 45(1): 295–301.

65. Choi S, Lee Y, Joo C, Lee S, Park J, Han K. Electrospun PVDF nanofiber web as polymer electrolyte or separator. *Electrochimica Acta.* 2004; 50(2–3): 339–343.

66. Wang J, Dai L, Gao Q, Wu P, Wang X. Electrospun nanofibers of polyferrocenylsilanes with different substituents at silicon. *European Polymer Journal.* 2008; 44(3): 602–607.

67. Lee K, Ohsawa O, Watanabe K, Kim I, Givens S, Chase B, et al. Electrospinning of Syndiotactic Polypropylene from a Polymer Solution at Ambient Temperatures. *Macromolecules.* 2009; 42(14): 5215–5218.

68. Kongkhlang T, Kotaki M, Kousaka Y, Umemura T, Nakaya D, Chirachanchai S. Electrospun polyoxymethylene: spinning conditions and its consequent nanoporous nanofiber. *Macromolecules.* 2008; 41(13): 4746–4752.

69. Chronakis I, Grapenson S, Jakob A. Conductive polypyrrole nanofibers via electrospinning: electrical and morphological properties. *Polymer.* 2006; 47(5): 1597–1603.

70. Electrospinning Nanofibers of Polyaniline and Polyaniline / (Polystyrene and Polyethylene Oxide) Blends.

71. Manandhar S, Vidhate S, D'Souza N. Water soluble levan polysaccharide biopolymer electrospun fibers. *Carbohydrate Polymers.* 2009; 78(4): 794–798.

72. Zhao N, Shi S, Lu G, Wei M. Polylactide (PLA)/layered double hydroxides composite fibers by electrospinning method. *Journal of Physics and chemistry of solids.* 2008; 69(5–6): 1564–1568.

73. Sanders EH, Kloefkorn R, Bowlin GL, Simpson DG, Wnek GE. Two-phase electrospinning from a single electrified jet: microencapsulation of aqueous reservoirs in poly(ethylene-co-vinyl acetate) fibers. *Macromolecules.* 2003; 36(11): 3803–3805.

74. Piperno S, Lozzi L, Rastelli R, Passacantando M, Santucci S. PMMA nanofibers production by electrospinning. *Applied Surface Science.* 2006; 252(15): 5583–5586.

75. Lyoo W, Lee K, Lee Y, Kim H. Preparation of high molecular weight poly (vinyl carbazole) web by electrospinning. In *Experimental Analysis of Nano and Engineering Materials and Structures* (Springer);. pp. 47–48.

76. Mit-uppatham C, Nithitanakul M, Supaphol P. Ultrafine electrospun polyamide-6 fibers: effect of solution conditions on morphology and average fiber diameter. *Macromolecular Chemistry and Physics.* 2004; 205(17): 2327–2338.

77. Xin X, Hussain M, Mao JJ. Continuing differentiation of human mesenchymal stem cells and induced chondrogenic and osteogenic lineages in electrospun PLGA nanofiber scaffold. *Biomaterials.* 2007; 28(2): 316–325.

78. Kim SH, Kim S-H, Nair S, Moore E. Reactive electrospinning of cross-linked poly(2-hydroxyethyl methacrylate) nanofibers and elastic properties of individual hydrogel nanofibers in aqueous solutions. *Macromolecules.* 2005; 38(9): 3719–3723.

79. Choi S, Lee S, Joo C, Im S, Kim S. Formation of interfiber bonding in electrospun poly (etherimide) nanofiber web. *Journal of Materials Science.* 2004; 39(4): 1511–1513.

80. Chen Z, Foster M, Zhou W, Fong H, Reneker D, Resendes R, et al. Structure of poly (ferrocenyldimethylsilane) in electrospun nanofibers. *Macromolecules.* 2001; 34(18): 6156–6158.

81. Kenawy E, Layman J, Watkins J, Bowlin G, Matthews J, Simpson D, et al. Electrospinning of poly (ethylene-co-vinyl alcohol) fibers. *Biomaterials.* 2003; 24(6): 907–913.

82. Yang W, Yu H, Zhu M, Bai H, Chen Y. Poly (m-Phenylene Isophthalamide) ultrafine fibers from an ionic liquid solution by dry-jet-wet-electrospinning. *Journal of Macromolecular Science, Part B.* 2006; 45(4): 573–579.

83. Hench LL, West JK. The sol-gel process. *Chemical Reviews* (Washington, DC, United States). 1990; 90(1): 33–72.

84. Hsu C-M, Shivkumar S. Nano-sized beads and porous fiber constructs of poly(.vepsiln.-caprolactone) produced by electrospinning. *Journal of Materials Science.* 2004; 39(9): 3003–3013.

85. Casper CL, Stephens JS, Tassi NG, Chase DB, Rabolt JF. Controlling surface morphology of electrospun polystyrene fibers: effect of humidity and molecular weight in the electrospinning process. *Macromolecules.* 2004; 37(2): 573–578.

86. Kessick R, Fenn J, Tepper G. The use of AC potentials in electrospraying and electrospinning processes. *Polymer.* 2004; 45(9): 2981–2984.

87. Buchko CJ, Chen LC, Shen Y, Martin DC. Processing and microstructural characterization of porous biocompatible protein polymer thin films. *Polymer.* 1999; 40(26): 7397–7407.

88. Baker SC, Atkin N, Gunning PA, Granville N, Wilson K, Wilson D, et al. Characterisation of electrospun polystyrene scaffolds for three-dimensional in vitro biological studies. *Biomaterials.* 2006; 27(16): 3136–3146.

89. Megelski S, Stephens JS, Chase DB, Rabolt JF. Micro- and nanostructure surface morphology on electrospun polymer fibers. *Macromolecules.* 2002; 35(22): 8456–8466.

90. Chronakis IS. Novel nanocomposites and nanoceramics based on polymer nanofibers using electrospinning process—a review. *Journal of Materials Processing Technology.* 2005; 167(2–3): 283–293.

91. Zussman E, Yarin A, Bazilevsky A, Avrahami R, Feldman M. Electrospun polyacrylonitrile/poly (methyl methacrylate)-derived turbostratic carbon micro-/nanotubes. *Advanced Materials.* 2006; 18(3): 348–353.

92. McCann JT LD, Xia YN. Electrospinning of nanofibers with core-sheath, hollow, or porous structures. *Journal of Materials Chemistry.* 2005; 15(7): 735–738.

93. Moghe A, Gupta B. Co-axial electrospinning for nanofiber structures: preparation and applications. *Polymer Reviews.* 2008; 48(2): 353–377.

94. Reznik S, Yarin A, Zussman E, Bercovici L. Evolution of a compound droplet attached to a core-shell nozzle under the action of a strong electric field. *Physics of Fluids.* 2006; 18: 062101.

95. Wilkes GL, Gupta P. Some investigations on the fiber formation by utilizing a side-by-side bicomponent electrospinning approach. *Polymer.* 2003; 44(20): 6353–6359.

96. Bhaskar S, Lahann J. Microstructured materials based on multicompartmental fibers. *Journal of the American Chemical Society.* 2009; 131(19): 6650–6651.

97. Maheshwari S, Chang H. Assembly of multi-stranded nanofiber threads through AC electrospinning. *Advanced Materials.* 2009; 21: 349–354.

98. Lyons J. Melt-electrospinning of Tthermoplastic polymers: an experimental and theoretical analysis. 2004.

99. OgataN, Shimada N, Yamaguchi S, Nakane K, Ogihara T. Melt-electrospinning of poly(ethylene terephthalate) and polyalirate. *Journal of Applied Polymer Science.* 2007; 105(3): 1127–1132.

100. Deng R, Liu Y, Ding Y, Xie P, Luo L, Yang W. Melt electrospinning of low-density polyethylene having a low-melt flow index. *Journal of Applied Polymer Science.* 2009; 114(1): 166–175.

101. Dalton P D, Lleixà Calvet J, Mourran A, Klee D, Möller M. Melt electrospinning of poly-(ethylene glycol-block-epsilon-caprolactone). *Biotechnology Journal.* 2006; 1(9): 998–1006.

102. Theron A Zussman E, Yarin AL. Electrostatic field-assisted alignment of electrospun nanofibres. *Nanotechnology.* 2001; 12(3): 384–390.

103. Katta P, Alessandro M, Ramsier RD, Chase GG. Continuous electrospinning of aligned polymer nanofibers onto a wire drum collector. *Nano Letters.* 2004; 4(11): 2215–2218.

104. Li D, Wang Y, Xia Y. Electrospinning of polymeric and ceramic nanofibers as uniaxially aligned arrays. *Nano Letters.* 2003; 3(8): 1167–1171.

105. Li D, Wang Y, Xia Y,. Electrospinning nanofibers as uniaxially aligned arrays and layer-by-layer stacked films. *Advanced Materials.* 2004; 16(4): 361–366.

106. Yang D, Lu B, Zhao Y, Jiang X. Fabrication of aligned fibrous arrays by magnetic electrospinning. *Advanced Materials.* 2007; 19(21): 3702–3706.

107. Carnell LS, Siochi EJ, Holloway NM, Stephens RM, Rhim C, Niklason LE, et al. Aligned mats from electrospun single fibers. *Macromolecules.* 2008; 41(14): 5345–5349.

108. Hellmann C, Belardi J, Dersch R, Greiner A, Wendorff JH, Bahnmueller S. High precision deposition electrospinning of nanofibers and nanofiber nonwovens. *Polymer.* 2009; 50(5): 1197–1205.

109. Zhang DM, Chang J. Electrospinning of three-dimensional nanofibrous tubes with controllable architectures. *Nano Letters.* 2008; 8(10): 3283–3287.

110. Smit E, Buttner U, Sanderson RD. Continuous yarns from electrospun fibers. *Polymer.* 2005; 46(8): 2419–2423.

111. Paul D. Dalton DK, Möller M. Electrospinning with dual collection rings. *Polymer.* 2005; 46(3): 611–614.

112. Huan Pan LL, Long Hu, Xiaojie Cui. Continuous aligned polymer fibers produced by a modified electrospinning method. *Polymer.* 2006; 47: 4901–4904.

113. Li D, Xia Y. Electrospinning of nanofibers: reinventing the wheel? *Advanced Materials.* 2004; 16(14): 1151–1170.

114. Greiner A, wendorff JH. Electrospinning: a fascinating method for the preparation of ultrathin fibers. *Angewandte Chemie, International Edition.* 2007; 46: 5670–5703.

115. Thavasi V, Singh G, Ramakrishna S. Electrospun nanofibers in energy and environmental applications. *Energy and Environmental Science.* 2008; 1(2): 205–221.

116. Fang J, Niu H, Lin T, Wang X. Applications of electrospun nanofibers. *Chinese Science Bulletin.* 2008; 53(15): 2265–2286.

117. Yoo HS, Kim TG, Park TG. Surface-functionalized electrospun nanofibers for tissue engineering and drug delivery. *Advanced Drug Delivery Reviews.* 2009; 61(12): 1033–1042.

118. Cui W, Zhou Y, Chang J. Electrospun nanofibrous materials for tissue engineering and drug delivery. *Science and Technology for Advanced Materials.* 2010; 11(1): doi:10.1088/1468-6996/11/1/014108.

119. Ma PX. Tissue engineering. In *Encyclopedia of Polymer Science and Technology.* NJ: John Wiley & Sons; 2004.

120. Xu C, Yang F, Wang S, Ramakrishna S. In vitro study of human vascular endothelial cell function on materials with various surface roughness. *Journal of Biomedical Materials Research, Part A.* 2004; 71(1): 154–161.

121. Mo XM, Xu CY, Kotaki M, Ramakrishna S. Electrospun P(LLA-CL) nanofiber: a biomimetic extracellular matrix for smooth muscle cell and endothelial cell proliferation. *Biomaterials.* 2004; 25(10): 1883–1890.

122. Rubenstein D, Han D, Goldgraben S, El-Gendi H, Gouma P-I, Frame MD. Bioassay chamber for angiogenesis with perfused explanted arteries and electrospun scaffolding. *Microcirculation.* 2007; 14(7): 723–737.

123. Balguid A, Mol A, van MMH, Bank RA, Bouten CV, Baaijens FP. Tailoring fiber diameter in electrospun poly(epsilon-caprolactone) scaffolds for optimal cellular infiltration in cardiovascular tissue engineering. *Tissue Engineering, Part A.* 2009; 15(2): 437–444.

124. Vaz CM, Van Tuijl S, Bouten CV, Baaijens FP. Design of scaffolds for blood vessel tissue engineering using a multi-layering electrospinning technique. *Acta Biomaterialia.* 2005; 1(5): 575–582.

125. Soletti L, Hong Y, Guan J, Stankus JJ, El-Kurdi MS, Wagner WR, et al. A bilayered elastomeric scaffold for tissue engineering of small diameter vascular grafts. *Acta Biomaterialia.* 2010; 6(1): 110–122.

126. Ma Z, He W, Yong T, Ramakrishna S. Grafting of gelatin on electrospun poly(caprolactone) nanofibers to improve endothelial cell spreading

and proliferation and to control cell orientation. *Tissue Engineering.* 2005; 11(7/8): 1149–1158.

127. Casper CL, Yang W, Farach-Carson MC, Rabolt JF. Coating electrospun collagen and gelatin fibers with perlecan domain I for increased growth factor binding. *Biomacromolecules.* 2007; 8(4): 1116–1123.

128. Zhu Y, Leong Meng F, Ong Wey F, Chan Park Mary B, Chian Kerm S. Esophageal epithelium regeneration on fibronectin grafted poly(L-lactide-co-caprolactone) (PLLC) nanofiber scaffold. *Biomaterials.* 2007; 28(5): 861–868.

129. Sell SA, McClure MJ, Barnes CP, Knapp DC, Walpoth BH, Simpson DG, et al. Electrospun polydioxanone-elastin blends: potential for bioresorbable vascular grafts. *Biomedical Materials.* 2006; 1(2): 72–80.

130. Lee SJ, Yoo JJ, Lim GJ, Atala A, Stitzel J. In vitro evaluation of electrospun nanofiber scaffolds for vascular graft application. *Journal of Biomedical Materials Research, Part A.* 2007; 83(4): 999–1008.

131. Telemeco TA, Ayres C, Bowlin GL, Wnek GE, Boland ED, Cohen N, et al. Regulation of cellular infiltration into tissue engineering scaffolds composed of submicron diameter fibrils produced by electrospinning. *Acta Biomaterialia.* 2005; 1(4): 377–385.

132. Tillman BW, Yazdani SK, Lee SJ, Geary RL, Atala A, Yoo JJ. The in vivo stability of electrospun polycaprolactone-collagen scaffolds in vascular reconstruction. *Biomaterials.* 2008; 30(4): 583–588.

133. Li WJ, Laurencin CT, Caterson EJ, Tuan RS, Ko FK. Electrospun nanofibrous structure: a novel scaffold for tissue engineering. *Journal of Biomedical Materials Research.* 2002; 60(4): 613–621.

134. Yoshimoto H, Shin YM, Terai H, Vacanti JP. A biodegradable nanofiber scaffold by electrospinning and its potential for bone tissue engineering. *Biomaterials.* 2003; 24(12): 2077–2082.

135. Shin M, Yoshimoto H, Vacanti JP. In vivo bone tissue engineering using Mesenchymal stem cells on a novel electrospun nanofibrous scaffold. *Tissue Engineering.* 2004; 10(1/2): 33–41.

136. Ekaputra AK, Zhou Y, Cool SM, Hutmacher DW. Composite electrospun scaffolds for engineering tubular bone grafts. *Tissue Engingeering, Part A.* 2009; 15(12): 3779–3788.

137. Zhang Y, Ouyang H, Lim CT, Ramakrishna S, Huang Z-M. Electrospinning of gelatin fibers and gelatin/PCL composite fibrous scaffolds. *Journal of Biomedical Materials Research, Part B: Applied Biomaterials.* 2005; 72(1): 156–165.

138. Fujihara K, Kotaki M, Ramakrishna S. Guided bone regeneration membrane made of polycaprolactone/calcium carbonate composite nano-fibers. *Biomaterials.* 2005; 26(19): 4139–4147.

139. Wutticharoenmongkol P, Sanchavanakit N, Pavasant P, Supaphol P. Preparation and characterization of novel bone scaffolds based on electrospun polycaprolactone fibers filled with nanoparticles. *Macromolecular Bioscience.* 2006; 6(1): 70–77.

140. Duan B, Wu L, Li X, Yuan X, Li X, Zhang Y, et al. Degradation of electrospun PLGA-chitosan/PVA membranes and their cytocompatibility in vitro. *Journal of Biomaterials Science, Polymer Edition.* 2007; 18(1): 95–115.

141. Tuzlakoglu K, Bolgen N, Salgado AJ, Gomes ME, Piskin E, Reis RL. Nano- and micro-fiber combined scaffolds: a new architecture for bone tissue engineering. *Journal of Materials Science: Materials in Medicine.* 2005; 16(12): 1099–1104.

142. Pham QP, Sharma U, Mikos AG. Electrospun poly(e-caprolactone) microfiber and multilayer nanofiber/microfiber scaffolds: characterization of scaffolds and measurement of cellular infiltration. *Biomacromolecules.* 2006; 7(10): 2796–2805.

143. Sisson K, Zhang C, Farach-Carson MC, Chase DB, Rabolt JF. Fiber diameters control osteoblastic cell migration and differentiation in electrospun gelatin. *Journal of Biomedical Materials Research, Part A.* 2010; 94(4): 1312–1320.

144. Binulal NS, Deepthy M, Selvamurugan N, Shalumon KT, Suja S, Mony U, et al. Role of nanofibrous poly(caprolactone) scaffolds in human Mesenchymal stem cell attachment and spreading for in vitro bone tissue engineering-response to osteogenic regulators. *Tissue Engeineering, Part A.* 2010; 16(2): 393–404.

145. Xu X, Zhuang X, Chen X, Wang X, Yang L, Jing X. Preparation of core-sheath composite nanofibers by emulsion electrospinning. *Macromolecular Rapid Communications.* 2006; 27(19): 1637–1642.

146. Kim HW, Lee HH, Chun GS. Bioactivity and osteoblast responses of novel biomedical nanocomposites of bioactive glass nanofiber filled poly(lactic acid). *Journal of Biomedical Materials Research, Part A.* 2008; 85A(3): 651–663.

147. Jo J, Lee E, Shin D, Kim H, Kim H, Koh Y, et al. In vitro/in vivo biocompatibility and mechanical properties of bioactive glass nanofiber and poly(ε-caprolactone) composite materials. *Journal of Biomedical Materials Research, Part B.* 2009; 91(1): 213–220.

148. Rath B, Nam J, Knobloch TJ, Lannutti JJ, Agarwal S. Compressive forces induce osteogenic gene expression in calvarial osteoblasts. *Journal of Biomechanics.* 2008; 41(5): 1095–1103.

149. Kannarkat JT, Battogtokh J, Philip J, Wilson OC, Mehl PM. Embedding of magnetic nanoparticles in polycaprolactone nanofiber scaffolds to facilitate bone healing and regeneration. *Journal of Applied Physics.* 2010; 107(9, Pt. 2): 09B307/1–09B/3.

150. Moroni L, Licht R, De Boer J, De Wijn JR, Van Blitterswijk CA. Fiber diameter and texture of electrospun PEOT/PBT scaffolds influence human mesenchymal stem cell proliferation and morphology, and the release of incorporated compounds. *Biomaterials.* 2006; 27(28): 4911–4922.

151. Ko EK, Jeong SI, Rim NG, Lee YM, Shin H, Lee B-K. In vitro osteogenic differentiation of human Mesenchymal stem cells and In vivo bone formation in composite nanofiber meshes. *Tissue Engineering, Part A.* 2008; 14(12): 2105–2119.

152. Schneider OD, Weber F, Brunner TJ, Loher S, Ehrbar M, Schmidlin PR, et al. In vivo and in vitro evaluation of flexible, cottonwool-like nanocomposites as bone substitute material for complex defects. *Acta Biomaterialia.* 2009; 5(5): 1775–1784.

153. Matthews JA, Boland ED, Wnek GE, Simpson DG, Bowlin GL. Electrospinning of collagen type II: a feasibility study. *Journal of Bioactive and Compatible Polymers.* 2003; 18(2): 125–134.

154. Shields KJ, Beckman MJ, Bowlin GL, Wayne JS. Mechanical properties and cellular proliferation of electrospun collagen type II. *Tissue Engineering.* 2004; 10(9/10): 1510–1517.

155. Li WJ, Danielson KG, Alexander PG, Tuan RS. Biological response of chondrocytes cultured in three-dimensional nanofibrous poly(ε-caprolactone) scaffolds. *Journal of Biomedical Materials Research, Part A.* 2003; 67(4): 1105–1114.

156. Lee IS, Kwon OH, Meng W, Kang IK, Ito Y. Nanofabrication of microbial polyester by electrospinning promotes cell attachment. *Macromolecular Research.* 2004; 12(4): 374–378.

157. Kwon OH, Lee IS, Ko YG, Meng W, Jung KH, Kang IK, et al. Electrospinning of microbial polyester for cell culture. *Biomedical Materials.* 2007; 2(1): S52–S58.

158. Bhattarai N, Edmondson D, Veiseh O, Matsen FA, Zhang M. Electrospun chitosan-based nanofibers and their cellular compatibility. *Biomaterials.* 2005; 26(31): 6176–6184.

159. Chen J, Li S, Chiang Y. Bioactive collagen-grafted poly-L-lactic acid nanofibrous membrane for cartilage tissue engineering. *Journal of Nanoscience and Nanotechnology.* 2010; 10(8): 5393–5398.

160. Chen J, Su C. Surface modification of electrospun PLLA nanofibers by plasma treatment and cationized gelatin immobilization for cartilage tissue engineering. *Acta Biomaterialia.* 2011; 7(1): 234–243.

161. Shanmugasundaram S, Chaudhry H, Arinzeh TL. Microscale versus nanoscale scaffold architecture for Mesenchymal stem cell chondrogenesis. *Tissue Engineering, Part A.* 2011; 17(5–6): 831–840.

162. Moroni L, Schotel R, Hamann D, de WJR, van BCA. 3D fiber-deposited electrospun integrated scaffolds enhance cartilage tissue formation. *Advanced Functional Materials.* 2008; 18(1): 53–60.

163. Thorvaldsson A, Stenhamre H, Gatenholm P, Walkenstroem P. Electrospinning of highly porous scaffolds for cartilage regeneration. *Biomacromolecules.* 2008; 9(3): 1044–1049.

164. Shim IK, Suh WH, Lee SY, Lee SH, Heo SJ, Lee MC, et al. Chitosan nano-/microfibrous double-layered membrane with rolled-up three-dimensional structures for chondrocyte cultivation. *Journal of Biomedical Materials Research, Part A.* 2009; 90(2): 595–602.

165. Stankus JJ, Guan J, Wagner WR. Fabrication of biodegradable elastomeric scaffolds with sub-micron morphologies. *Journal of Biomedical Materials Research, Part A.* 2004; 70(4): 603–614.

166. Venugopal J, Zhang YZ, Ramakrishna S. Fabrication of modified and functionalized polycaprolactone nanofibre scaffolds for vascular tissue engineering. *Nanotechnology.* 2005; 16(10): 2138–2142.

167. Chen ZG, Wang PW, Wei B, Mo XM, Cui FZ. Electrospun collagen-chitosan nanofiber: a biomimetic extracellular matrix for endothelial cell and smooth muscle cell. *Acta Biomaterialia.* 2010; 6(2): 372–382.

168. Riboldi SA, Sampaolesi M, Neuenschwander P, Cossu G, Mantero S. Electrospun degradable polyesterurethane membranes: potential scaffolds for skeletal muscle tissue engineering. *Biomaterials.* 2005; 26(22): 4606–4615.

169. Riboldi SA, Sadr N, Pigini L, Neuenschwander P, Simonet M, Mognol P, et al. Skeletal myogenesis on highly orientated microfibrous polyesterurethane scaffolds. *Journal of Biomedical Materials Research, Part A.* 2008; 84(4): 1094–1101.

170. Dong Y, Yong T, Liao S, Chan CK, Stevens MM, Ramakrishna S. Distinctive degradation behaviors of electrospun polyglycolide, poly(L-lactide-co-

glycolide), and poly(L-lactide-co-ε-caprolactone) nanofibers cultured with/without porcine smooth muscle cells. *Tissue Engineering, Part A.* 2010; 16(1): 283–298.

171. Min BM, Lee G, Kim SH, Nam YS, Lee TS, Park WH. Electrospinning of silk fibroin nanofibers and its effect on the adhesion and spreading of normal human keratinocytes and fibroblasts in vitro. *Biomaterials.* 2003; 25(7–8): 1289–1297.

172. Rho KS, Jeong L, Lee G, Seo BM, Park YJ, Hong SD, et al. Electrospinning of collagen nanofibers: effects on the behavior of normal human keratinocytes and early-stage wound healing. *Biomaterials.* 2006; 27(8): 1452–1461.

173. Zhang YZ, Venugopal J, Huang ZM, Lim CT, Ramakrishna S. Crosslinking of the electrospun gelatin nanofibers. *Polymer.* 2006; 47(8): 2911–2917.

174. Ji Y, Ghosh K, Shu XZ, Li B, Sokolov JC, Prestwich GD, et al. Electrospun three-dimensional hyaluronic acid nanofibrous scaffolds. *Biomaterials.* 2006; 27(20): 3782–3792.

175. Zhang YZ, Venugopal J, Huang ZM, Lim CT, Ramakrishna S. Characterization of the surface biocompatibility of the electrospun PCL-collagen nanofibers using fibroblasts. *Biomacromolecules.* 2005; 6(5): 2583–2589.

176. Bhattarai SR, Bhattarai N, Viswanathamurthi P, Yi HK, Hwang PH, Kim HY. Hydrophilic nanofibrous structure of polylactide; fabrication and cell affinity. *Journal of Biomedical Materials Research, Part A.* 2006; 78(2): 247–257.

177. Babaeijandaghi F, Shabani I, Seyedjafari E, Naraghi ZS, Vasei M, Haddadi-Asl V, et al. Accelerated epidermal regeneration and improved dermal reconstruction achieved by polyethersulfone nanofibers. *Tissue Engineering, Part A.* 2010; 16(11): 3527–3536.

178. Zhao P, Jiang H, Pan H, Zhu K, Chen W. Biodegradable fibrous scaffolds composed of gelatin coated poly(ε-caprolactone) prepared by coaxial electrospinning. *Journal of Biomedical Materials Research, Part A.* 2007; 83(2): 372–382.

179. Duan B, Yuan X, Zhu Y, Zhang Y, Li X, Zhang Y, et al. A nanofibrous composite membrane of PLGA-chitosan/PVA prepared by electrospinning. *European Polymer Journal.* 2006; 42(9): 2013–2022.

180. Park K, Jung HJ, Kim JJ, Ahn KD, Han DK, Ju YM. Acrylic acid-grafted hydrophilic electrospun nanofibrous poly(L-lactic acid) scaffold. *Macromolecular Research.* 2006; 14(5): 552–558.

181. Park K, Ju YM, Son JS, Ahn KD, Han DK. Surface modification of biodegradable electrospun nanofiber scaffolds and their interaction with fibroblasts. *Journal of Biomaterials Science, Polymer Edition.* 2007; 18(4): 369–382.

182. Powell HM, Boyce ST. Engineered human skin fabricated using electrospun collagen-PCL blends: morphogenesis and mechanical properties. *Tissue Engineering, Part A.* 2009; 15(8): 2177–2187.

183. Kim TG, Park TG. Biomimicking extracellular matrix: cell adhesive RGD peptide modified electrospun poly(D,L-lactic-co-glycolic acid) nanofiber mesh. *Tissue Engineering.* 2006; 12(2): 221–233.

184. Zhong S, Teo W, Zhu X, Beuerman RW, Ramakrishna S, Yung LYL. An aligned nanofibrous collagen scaffold by electrospinning and its effects on in vitro fibroblast culture. *Journal of Biomedical Materials Research, Part A.* 2006; 79(3): 456–463.

185. Kurpinski KT, Stephenson JT, Janairo RRR, Lee H, Li S. The effect of fiber alignment and heparin coating on cell infiltration into nanofibrous PLLA scaffolds. *Biomaterials.* 2010; 31(13): 3536–3542.

186. Ki CS, Kim JW, Hyun JH, Lee KH, Hattori M, Rah DK, et al. Electrospun three-dimensional silk fibroin nanofibrous scaffold. *Journal of Applied Polymer Science.* 2007; 106(6): 3922–3928.

187. Kim G, Kim W. Highly porous 3D nanofiber scaffold using an electrospinning technique. *Journal of Biomedical Materials Research, Part B: Applied Biomaterials.* 2007; 81(1): 104–110.

188. Sun T, Norton D, McKean RJ, Haycock JW, Ryan AJ, MacNeil S. Development of a 3D cell culture system for investigating cell interactions with electrospun fibers. *Biotechnology and Bioengineering.* 2007; 97(5): 1318–1328.

189. Powell HM, Boyce ST. Fiber density of electrospun gelatin scaffolds regulates morphogenesis of dermal-epidermal skin substitutes. *Journal of Biomedical Materials Research, Part A.* 2008; 84(4): 1078–1086.

190. Yang F, Xu CY, Kotaki M, Wang S, Ramakrishna S. Characterization of neural stem cells on electrospun poly(L-lactic acid) nanofibrous scaffold. *Journal of Biomaterials Science, Polymer Edition.* 2004; 15(12): 1483–1497.

191. Yang F, Murugan R, Wang S, Ramakrishna S. Electrospinning of nano/micro scale poly(L-lactic acid) aligned fibers and their potential in neural tissue engineering. *Biomaterials.* 2005; 26(15): 2603–2610.

192. Mukhatyar VJ, Salmeron-Sanchez M, Rudra S, Mukhopadaya S, Barker TH, Garcia AJ, et al. Role of fibronectin in topographical guidance of

neurite extension on electrospun fibers. *Biomaterials.* 2011; 32(16): 3958–3968.

193. Christopherson GT, Song H, Mao H-Q. The influence of fiber diameter of electrospun substrates on neural stem cell differentiation and proliferation. *Biomaterials.* 2008; 30(4): 556–564.
194. Wang HB, Mullins ME, Cregg JM, McCarthy CW, Gilbert RJ. Varying the diameter of aligned electrospun fibers alters neurite outgrowth and Schwann cell migration. *Acta Biomaterialia.* 2010; 6(8): 2970–2978.
195. Nisbet DR, Pattanawong S, Ritchie NE, Shen W, Finkelstein DI, Horne MK, et al. Interaction of embryonic cortical neurons on nanofibrous scaffolds for neural tissue engineering. *Journal of Neural Engineering.* 2007; 4(2): 35–41.
196. Nisbet DR, Yu LMY, Zahir T, Forsythe JS, Shoichet MS. Characterization of neural stem cells on electrospun poly(Îμ-caprolactone) submicron scaffolds: evaluating their potential in neural tissue engineering. *Journal of Biomaterials Science, Polymer Edition.* 2008; 19(5): 623–634.
197. Sangsanoh P, Waleetorncheepsawat S, Suwantong O, Wutticharoenmongkol P, Weeranantanapan O, Chuenjitbuntaworn B, et al. In vitro biocompatibility of Schwann cells on curfaces of biocompatible polymeric electrospun fibrous and solution-cast film scaffolds. *Biomacromolecules.* 2007; 8(5): 1587–1594.
198. Lee JY, Bashur CA, Goldstein AS, Schmidt CE. Polypyrrole-coated electrospun PLGA nanofibers for neural tissue applications. *Biomaterials.* 2009; 30(26): 4325–4335.
199. Xie J, MacEwan MR, Willerth SM, Li X, Moran DW, Sakiyama-Elbert SE, et al. Conductive core-sheath nanofibers and their potential application in neural tissue engineering. *Advanced Functional Materials.* 2009; 19(14): 2312–2318.
200. Bini TB, Gao S, Tan TC, Wang S, Lim A, Hai LB, et al. Electrospun poly(L-lactide-co-glycolide) biodegradable polymer nanofiber tubes for peripheral nerve regeneration. *Nanotechnology.* 2004; 15(11): 1459–1464.
201. Wang C, Zhang K, Fan C, Mo X, Ruan H, Li F. Aligned natural-synthetic polyblend nanofibers for peripheral nerve regeneration. *Acta Biomaterialia.* 2011; 7(2): 634–643.
202. Bini TB, Gao S, Wang S, Ramakrishna S. Poly(L-lactide-co-glycolide) biodegradable microfibers and electrospun nanofibers for nerve tissue engineering: an in vitro study. *Journal of Materials Science.* 2006; 41(19): 6453–6459.

203. Ito Y, Hasuda H, Kamitakahara M, Ohtsuki C, Tanihara M, Kang I-K, et al. A composite of hydroxyapatite with electrospun biodegradable nanofibers as a tissue engineering material. *Journal of Bioscience and Bioengineering.* 2005; 100(1): 43–49.

204. Lim SH, Mao HQ. Electrospun scaffolds for stem cell engineering. *Advanced Drug Delivery Reviews.* 2009; 61(12): 1084–1096.

205. Zong X, Bien H, Chung C, Yin L, Fang D, Hsiao BS, et al. Electrospun fine-textured scaffolds for heart tissue constructs. *Biomaterials.* 2005; 26(26): 5330–5338.

206. Lee CH, Shin HJ, Cho IH, Kang YM, Kim IA, Park KD, et al. Nanofiber alignment and direction of mechanical strain affect the ECM production of human ACL fibroblast. *Biomaterials.* 2005; 26(11): 1261–1270.

207. McManus M, Boland E, Sell S, Bowen W, Koo H, Simpson D, et al. Electrospun nanofibre fibrinogen for urinary tract tissue reconstruction. *Biomedical Materials.* 2007; 2(4): 257–262.

208. Marler JJ, Upton J, Langer R, Vacanti JP. Transplantation of cells in matrixes for tissue regeneration. *Advanced Drug Delivery Reviews.* 1998; 33(1–2): 165–182.

209. Khil MS, Cha DI, Kim HY, Kim IS, Bhattarai N. Electrospun nanofibrous polyurethane membrane as wound dressing. *Journal of Biomedical Materials Research, Part B: Applied Biomaterials.* 2003; 67(2): 675–679.

210. Schneider A, Wang XY, Kaplan DL, Garlick JA, Egles C. Biofunctionalized electrospun silk mats as a topical bioactive dressing for accelerated wound healing. *Acta Biomaterialia.* 2009; 5(7): 2570–2578.

211. Merrell JG, McLaughlin SW, Tie L, Laurencin CT, Chen AF, Nair LS. Curcumin-loaded poly(ε-caprolactone) nanofibres: diabetic wound dressing with anti-oxidant and anti-inflammatory properties. *Clinical and Experimental Pharmacology and Physiology.* 2009; 36(12): 1149–1156.

212. Suwantong O, Ruktanonchai U, Supaphol P. In vitro biological evaluation of electrospun cellulose acetate fiber mats containing asiaticoside or curcumin. *Journal of Biomedical Materials Research, Part A.* 2010; 94(4): 1216–1225.

213. Rho KS, Jeong L, Lee G, Seo BM, Park YJ, Hong SD, et al. Electrospinning of collagen nanofibers: effects on the behavior of normal human keratinocytes and early-stage wound healing. *Biomaterials.* 2006; 27(8): 1452–1461.

214. Liu SJ, Kau YC, Chou CY, Chen JK, Wu RC, Yeh WL. Electrospun PLGA/collagen nanofibrous membrane as early-stage wound dressing. *Journal of Membrane Science.* 2010; 355(1–2): 53–59.

215. Noh HK, Lee SW, Kim JM, Oh JE, Kim KH, Chung CP, et al. Electrospinning of chitin nanofibers: degradation behavior and cellular response to normal human keratinocytes and fibroblasts. *Biomaterials.* 2006; 27(21): 3934–3944.

216. Chong EJ, Phan TT, Lim IJ, Zhang YZ, Bay BH, Ramakrishna S, et al. Evaluation of electrospun PCL/gelatin nanofibrous scaffold for wound healing and layered dermal reconstitution. *Acta biomaterialia.* 2007; 3(3): 321–330.

217. Choi JS, Leong KW, Yoo HS. In vivo wound healing of diabetic ulcers using electrospun nanofibers immobilized with human epidermal growth factor (EGF). *Biomaterials.* 2008; 29(5): 587–596.

218. Liu X, Lin T, Fang J, Yao G, Wang X. Electrospun nanofibre membranes as wound dressing materials. *Advances in Science and Technology. 2008;57 (Biomedical Applications of Smart Materials, Nanotechnology and Micro/Nano Engineering)*: 125–130.

219. Liu X, Lin T, Fang J, Yao G, Zhao H, Dodson M, et al. In vivo wound healing and antibacterial performances of electrospun nanofibre membranes. *Journal of Biomedical Materials Research, Part A.* 2010; 94(2): 499–508.

220. Menzies D. Peritoneal adhesions. Incidence, cause, and prevention. *Surgery Annual.* 1992; 24 Pt 1: 27–45.

221. DeCherney AH, diZerega GS. Clinical problem of intraperitoneal postsurgical adhesion formation following general surgery and the use of adhesion prevention barriers. *The Surgical Clinics of North America.* 1997; 77(3): 671–688.

222. Bolgen N, Vargel I, Korkusuz P, Menceloglu YZ, Piskin E. In vivo performance of antibiotic embedded electrospun PCL membranes for prevention of abdominal adhesions. *Journal of Biomedical Materials Research, Part B: Applied Biomaterials.* 2007; 81(2): 530–543.

223. Hong KH. Preparation and properties of electrospun poly(vinyl alcohol)/silver fiber web as wound dressings. *Polymer Engineering and Science.* 2007; 47(1): 43–49.

224. Lala NL, Ramaseshan R, Li B, Sundarrajan S, Barhate RS, Liu Y-j, et al. Fabrication of nanofibers with antimicrobial functionality used as filters: protection against bacterial contaminants. *Biotechnology and Bioengineering.* 2007; 97(6): 1357–1365.

225. Wu J, Hou S, Ren D, Mather PT. Antimicrobial properties of nanostructured hydrogel webs containing silver. *Biomacromolecules.* 2009; 10(9): 2686–2693.

226. Chen J, Chiang Y. Bioactive electrospun silver nanoparticles-containing polyurethane nanofibers as wound dressings. *Journal of Nanoscience and Nanotechnology.* 2010; 10(11): 7560–7564.

227. Ignatova M, Manolova N, Rashkov I. Electrospinning of poly(vinyl pyrrolidone)-iodine complex and poly(ethylene oxide)/poly(vinyl pyrrolidone)-iodine complex—a prospective route to antimicrobial wound dressing materials. *European Polymer Journal.* 2007; 43(5): 1609–1623.

228. Spadaro JA, Berger TJ, Barranco SD, Chapin SE, Becker RO. Antibacterial effects of silver electrodes with weak direct current. *Antimicrobial Agents and Chemotherapy.* 1974; 6(5): 637–642.

229. Kenawy E-R, Bowlin GL, Mansfield K, Layman J, Simpson DG, Sanders EH, et al. Release of tetracycline hydrochloride from electrospun poly(ethylene-co-vinylacetate), poly(lactic acid), and a blend. *Journal of Controlled Release.* 2002; 81(1–2): 57–64.

230. Vargas E, Do V, De B, De Q. Hyperbranched polyglycerol electrospun nanofibers for wound dressing applications. *Acta Biomaterialia.* 2010; 6(3): 1069–1078.

231. Yih TC, Al-Fandi M. Engineered nanoparticles as precise drug delivery systems. *Journal of Cellular Biochemistry.* 2006; 97(6): 1184–1190.

232. Zong X, Kim K, Fang D, Ran S, Hsiao BS, Chu B. Structure and process relationship of electrospun bioabsorbable nanofiber membranes. *Polymer.* 2002; 43(16): 4403–4412.

233. Verreck G, Chun I, Rosenblatt J, Peeters J, Van Dijck A, Mensch J, et al. Incorporation of drugs in an amorphous state into electrospun nanofibers composed of a water-insoluble, nonbiodegradable polymer. *Journal of Controlled Release.* 2003; 92(3): 349–360.

234. Verreck G, Chun I, Peeters J, Rosenblatt J, Brewster ME. Preparation and characterization of nanofibers containing amorphous drug dispersions generated by electrostatic spinning. *Pharmaceutical Research.* 2003; 20(5): 810–817.

235. Zeng J, Xu X, Chen X, Liang Q, Bian X, Yang L, et al. Biodegradable electrospun fibers for drug delivery. *Journal of Controlled Release.* 2003; 92(3): 227–231.

236. Luu YK, Kim K, Hsiao BS, Chu B, Hadjiargyrou M. Development of a nanostructured DNA delivery scaffold via electrospinning of PLGA and

PLA-PEG block copolymers. *Journal of Controlled Release.* 2003; 89(2): 341–353.

237. Kim HS, Yoo HS. MMPs-responsive release of DNA from electrospun nanofibrous matrix for local gene therapy: In vitro and in vivo evaluation. *Journal of Controlled Release.* 2010; 145(3): 264–271.

238. Zhang J, Duan Y, Wei D, Wang L, Wang H, Gu Z, et al. Co-electrospun fibrous scaffold-adsorbed DNA for substrate-mediated gene delivery. *Journal of Biomedical Materials Research, Part A.* 2011; 96(1): 212–220.

239. Zeng J, Aigner A, Czubayko F, Kissel T, Wendorff JH, Greiner A. Poly(vinyl alcohol) nanofibers by electrospinning as a protein delivery system and the retardation of enzyme release by additional polymer coatings. *Biomacromolecules.* 2005; 6(3): 1484–1488.

240. Chew SY, Wen J, Yim EKF, Leong KW. Sustained release of proteins from electrospun biodegradable fibers. *Biomacromolecules.* 2005; 6(4): 2017–2024.

241. Kim K, Luu YK, Chang C, Fang D, Hsiao BS, Chu B, et al. Incorporation and controlled release of a hydrophilic antibiotic using poly(lactide-co-glycolide)-based electrospun nanofibrous scaffolds. *Journal of Controlled Release.* 2004; 98(1): 47–56.

242. Yu DG, Shen XX, Branford-White C, White K, Zhu LM, Bligh SWA. Oral fast-dissolving drug delivery membranes prepared from electrospun polyvinylpyrrolidone ultrafine fibers. *Nanotechnology.* 2009; 20(5): 055104/1-/9.

243. Zeng J, Yang L, Liang Q, Zhang X, Guan H, Xu X, et al. Influence of the drug compatibility with polymer solution on the release kinetics of electrospun fiber formulation. *Journal of Controlled Release.* 2005; 105(1–2): 43–51.

244. Jiang H, Fang D, Hsiao B, Chu B, Chen W. Preparation and characterization of ibuprofen-loaded poly(lactide-co-glycolide)/poly(ethylene glycol)-g-chitosan electrospun membranes. *Journal of Biomaterials Science, Polymer Edition.* 2004; 15(3): 279–296.

245. Yang D, Li Y, Nie J. Preparation of gelatin/PVA nanofibers and their potential application in controlled release of drugs. *Carbohydrate Polymers.* 2007; 69(3): 538–543.

246. Kenawy ER, Abdel-Hay FI, El-Newehy MH, Wnek GE. Controlled release of ketoprofen from electrospun poly(vinyl alcohol) nanofibers. *Materials Science and Engineering, A: Structural Materials: Properties, Microstructure and Processing.* 2007; 459(1–2): 390–396.

247. Xu X, Chen X, Wang Z, Jing X. Ultrafine PEG-PLA fibers loaded with both paclitaxel and doxorubicin hydrochloride and their in vitro cytotoxicity. European *Journal of Pharmaceutics and Biopharmaceutics.* 2009; 72(1): 18–25.

248. Xu X, Yang L, Xu X, Wang X, Chen X, Liang Q, et al. Ultrafine medicated fibers electrospun from W/O emulsions. *Journal of Controlled Release.* 2005; 108(1): 33–42.

249. Im JS, Yun J, Lim Y, Kim H, Lee Y. Fluorination of electrospun hydrogel fibers for a controlled release drug delivery system. *Acta Biomaterialia.* 2010; 6(1): 102–109.

250. Jiang H, Hu Y, Li Y, Zhao P, Zhu K, Chen W. A facile technique to prepare biodegradable coaxial electrospun nanofibers for controlled release of bioactive agents. *Journal of Controlled Release.* 2005; 108(2–3): 237–243.

251. Jiang H, Hu Y, Zhao P, Li Y, Zhu K. Modulation of protein release from biodegradable core-shell structured fibers prepared by coaxial electrospinning. *Journal of Biomedical Materials Research, Part B: Applied Biomaterials.* 2006; 79(1): 50–57.

252. Huang Z-M, He C-L, Yang A, Zhang Y, Han X-J, Yin J, et al. Encapsulating drugs in biodegradable ultrafine fibers through co-axial electrospinning. *Journal of Biomedical Materials Research, Part A.* 2006; 77(1): 169–179.

253. Huang H, He C, Wang H, Mo X. Preparation of core-shell biodegradable microfibers for long-term drug delivery. *Journal of Biomedical Materials Research, Part A.* 2009; 90(4): 1243–1251.

254. He CL, Huang ZM, Han XJ. Fabrication of drug-loaded electrospun aligned fibrous threads for suture applications. *Journal of Biomedical Materials Research, Part A.* 2009; 89(1): 80–95.

255. Qi R, Guo R, Shen M, Cao X, Zhang L, Xu J, et al. Electrospun poly(lactic-co-glycolic acid)/halloysite nanotube composite nanofibers for drug encapsulation and sustained release. *Journal of Materials Chemistry.* 2010; 20(47): 10622–10629.

256. Kosmider K, Scott J. Polymeric nanofibres exhibit an enhanced air filtration performance. *Filtrationand Separation.* 2002; 39(6): 20–22.

257. Gibson P, Schreuder-Gibson H, Rivin D. Transport properties of porous membranes based on electrospun nanofibers. *Colloids and Surfaces, A: Physicochemical and Engineering Aspects.* 2001; 187–188: 469–481.

258. Barhate RS, Ramakrishna S. Nanofibrous filtering media: Filtration problems and solutions from tiny materials. *Journal of Membrane Science.* 2007; 296(1–2): 1–8.

259. Shin C, Chase GG, Reneker DH. Recycled expanded polystyrene nanofibers applied in filter media. *Colloids and Surfaces, A: Physicochemical and Engineering Aspects.* 2005; 262(1–3): 211–215.

260. Shin C, Chase GG, Reneker DH. The effect of nanofibers on liquid-liquid coalescence filter performance. *AIChE Journal.* 2005; 51(12): 3109–3113.

261. Yoon K, Kim K, Wang X, Fang D, Hsiao BS, Chu B. High flux ultrafiltration membranes based on electrospun nanofibrous PAN scaffolds and chitosan coating. *Polymer.* 2006; 47(7): 2434–2441.

262. Tang Z, Qiu C, McCutcheon JR, Yoon K, Ma H, Fang D, et al. Design and fabrication of electrospun polyethersulfone nanofibrous scaffold for high-flux nanofiltration membranes. *Journal of Polymer Science, Part B: Polymer Physics.* 2009; 47(22): 2288–2300.

263. Qin XH, Wang SY. Electrospun nanofibers from crosslinked poly(vinyl alcohol) and its filtration efficiency. *Journal of Applied Polymer Science.* 2008; 109(2): 951–956.

264. Homaeigohar SS, Buhr K, Ebert K. Polyethersulfone electrospun nanofibrous composite membrane for liquid filtration. *Journal of Membrane Science.* 2010; 365(1–2): 68–77.

265. Zhang Q, Welch J, Park H, Wu C, Sigmund W, Marijnissen JCM. Improvement in nanofiber filtration by multiple thin layers of nanofiber mats. *Journal of Aerosol Science.* 2010; 41(2): 230–236.

266. Barhate RS, Loong CK, Ramakrishna S. Preparation and characterization of nanofibrous filtering media. *Journal of Membrane Science.* 2006; 283: 209–218.

267. Gopal R, Kaur S, Ma Z, Chan C, Ramakrishna S, Matsuura T. Electrospun nanofibrous filtration membrane. *Journal of Membrane Science.* 2006; 281(1–2): 581–586.

268. Gopal R, Kaur S, Feng CY, Chand C, Ramakrishna S, Tabe S, et al. Electrospun nanofibrous polysulfone membranes as pre-filters: particulate removal. *Journal of Membrane Science.* 2007; 289: 210–219.

269. Bjorge D, Daels N, De VS, Dejans P, Van CT, Audenaert W, et al. Performance assessment of electrospun nanofibers for filter applications. *Desalination.* 2009; 249(3): 942–948.

270. Ki CS, Gang EH, Um IC, Park YH. Nanofibrous membrane of wool keratose/silk fibroin blend for heavy metal ion adsorption. *Journal of Membrane Science.* 2007; 302(1–2): 20–26.

271. Haider S, Park SY. Preparation of electrospun chitosan nanofibers and their applications to the adsorption of Cu(II) and Pb(II) ions from an

aqueous solution. *Journal of Membrane Science.* 2009; 328(1–2): 90–96.

272. Desai K, Kit K, Li J, Michael DP, Zivanovic S, Meyer H. Nanofibrous chitosan non-wovens for filtration applications. *Polymer.* 2009; 50(15): 3661–3669.

273. Kampalanonwat P, Supaphol P. Preparation and adsorption behavior of aminated electrospun polyacrylonitrile nanofiber mats for heavy metal ion removal. *ACS Applied Materials and Interfaces.* 2010; 2(12): 3619–3627.

274. Saeed K, Haider S, Oh TJ, Park SY. Preparation of amidoxime-modified polyacrylonitrile (PAN-oxime) nanofibers and their applications to metal ions adsorption. *Journal of Membrane Science.* 2008; 322(2): 400–405.

275. Hota G, Kumar BR, Ng WJ, Ramakrishna S. Fabrication and characterization of a boehmite nanoparticle impregnated electrospun fiber membrane for removal of metal ions. *Journal of Membrane Science.* 2008; 43(1): 212–217.

276. Wu S, Li F, Wang H, Fu L, Zhang B, Li G. Effects of poly(vinyl alcohol) (PVA) content on preparation of novel thiol-functionalized mesoporous PVA/SiO2 composite nanofiber membranes and their application for adsorption of heavy metal ions from aqueous solution. *Polymer.* 2010; 51(26): 6203–6211.

277. Wang H, Ding J, Lee B, Wang X, Lin T. Polypyrrole-coated electrospun nanofibre membranes for recovery of Au(III) from aqueous solution. *Journal of Membrane Science.* 2007; 303(1–2): 119–125.

278. Demir MM, Gulgun MA, Menceloglu YZ, Erman B, Abramchuk SS, Makhaeva EE, et al. Palladium nanoparticles by electrospinning from poly(acrylonitrile-co-acrylic acid)-$PdCl_2$ solutions. Relations between preparation conditions, particle size, and catalytic activity. *Macromolecules.* 2004; 37(5): 1787–1792.

279. Yu J, Liu T. Preparation of nano-fiber supported palladium catalysts and their use for the catalytic hydrogenation of olefins. *Acta Polymerica Sinica.* 2007; (6): 514–518.

280. Stasiak M, Studer A, Greiner A, Wendorff JH. Polymer fibers as carriers for homogeneous catalysts. *Chemistry--A European Journal.* 2007; 13(21): 6150–6156.

281. Chen L, Bromberg L, Hatton TA, Rutledge GC. Catalytic hydrolysis of p-nitrophenyl acetate by electrospun polyacrylamidoxime nanofibers. *Polymer.* 2007; 48(16): 4675–4682.

282. He C-H, Gong J. The preparation of PVA-Pt/TiO_2 composite nanofiber aggregate and the photocatalytic degradation of solid-phase polyvinyl alcohol. *Polymer Degradation and Stability.* 2003; 81(1): 117–124.

283. Choi SK, Kim S, Lim SK, Park H. Photocatalytic comparison of TiO_2 nanoparticles and electrospun TiO_2 nanofibers: effects of mesoporosity and interparticle charge transfer. *Journal of Physical Chemistry C.* 2010; 114(39): 16475–16480.

284. Li H, Zhang W, Li B, Pan W. Diameter-dependent photocatalytic activity of electrospun TiO_2 nanofiber. *Journal of the American Ceramic Society.* 2010; 93(9): 2503–2506.

285. Formo E, Yavuz MS, Lee EP, Lane L, Xia Y. Functionalization of electrospun ceramic nanofibre membranes with noble-metal nanostructures for catalytic applications. *Journal of Materials Chemistry.* 2009; 19(23): 3878–3882.

286. Pan C, Dong L. Fabrication of gold-doped titanium dioxide (TiO_2:Au) nanofibers photocatalyst by vacuum ion sputter coating. *Journal of Macromolecular Science, Part B: Physics.* 2009; 48(5): 919–926.

287. Wang C, Shao C, Zhang X, Liu Y. SnO_2 nanostructures-TiO_2 nanofibers heterostructures: controlled fabrication and high photocatalytic properties. *Inorganic Chemistry.* 2009; 48(15): 7261–7268.

288. Zhang Z, Shao C, Zhang L, Li X, Liu Y. Electrospun nanofibers of V-doped TiO_2 with high photocatalytic activity. *Journal of Colloid and Interface Science.* 2010; 351(1): 57–62.

289. Zhan S, Yang J, Liu Y, Wang N, Dai J, Yu H, et al. Mesoporous Fe_2O_3-doped TiO_2 nanostructured fibers with higher photocatalytic activity. *Journal of Colloid and Interface Science.* 2011; 355(2): 328–333.

290. Lin D, Wu H, Zhang R, Pan W. Enhanced photocatalysis of electrospun Ag-ZnO heterostructured nanofibers. *Chemistry of Materials.* 2009; 21(15): 3479–3484.

291. Dong B, Li Z, Li Z, Xu X, Song M, Zheng W, et al. Highly efficient $LaCoO_3$ nanofibers catalysts for photocatalytic degradation of Rhodamine B. *Journal of the American Ceramic Society.* 2010; 93(11): 3587–3590.

292. Sugunan A, Guduru VK, Uheida A, Toprak MS, Muhammed M. Radially oriented ZnO nanowires on flexible poly-L-lactide nanofibers for continuous-flow photocatalytic water purification. *Journal of the American Ceramic Society.* 2010; 93(11): 3740–3744.

293. Matatov-Meytal Y, Sheintuch M. Catalytic fibers and cloths. *Applied Catalysis A: General.* 2002; 231(1–2): 1–16.

294. Wang Y, Hsieh Y. Enzyme immobilization to ultra-fine cellulose fibers via amphiphilic polyethylene glycol spacers. *Journal of Polymer Science, Part A: Polymer Chemistry.* 2004; 42(17): 4289–4299.

295. Lee KH, Ki CS, Baek DH, Kang GD, Ihm DW, Park YH. Application of electrospun silk fibroin nanofibers as an immobilization support of enzyme. *Fibers and Polymers.* 2005; 6(3): 181–185.

296. Wang Z, Xu Z, Wan L, Wu J, Innocent C, Seta P. Nanofibrous membranes containing carbon nanotubes: electrospun for redox enzyme immobilization. *Macromolecular Rapid Communications.* 2006; 27(7): 516–521.

297. Li S, Chen J, Wu W. Electrospun polyacrylonitrile nanofibrous membranes for lipase immobilization. *Journal of Molecular Catalysis B: Enzymatic.* 2007; 47(3–4): 117–124.

298. Huang X, Ge D, Xu Z. Preparation and characterization of stable chitosan nanofibrous membrane for lipase immobilization. *European Polymer Journal.* 2007; 43(9): 3710–3718.

299. Stoilova O, Ignatova M, Manolova N, Godjevargova T, Mita DG, Rashkov I. Functionalized electrospun mats from styrene-maleic anhydride copolymers for immobilization of acetylcholinesterase. *European Polymer Journal.* 2010; 46(10): 1966–1974.

300. Jia H, Zhu G, Vugrinovich B, Kataphinan W, Reneker DH, Wang P. Enzyme-carrying polymeric nanofibers prepared via electrospinning for use as unique biocatalysts. *Biotechnology Progress.* 2002; 18(5): 1027–1032.

301. Kim BC, Nair S, Kim J, Kwak JH, Grate JW, Kim SH, et al. Preparation of biocatalytic nanofibers with high activity and stability via enzyme aggregate coating on polymer nanofibers. *Nanotechnology.* 2005; 16(7): 382–388.

302. Kim TG, Park TG. Surface functionalized electrospun biodegradable nanofibers for immobilization of bioactive molecules. *Biotechnology Progress.* 2006; 22(4): 1108–1113.

303. Stoilova O, Manolova N, Gabrovska K, Marinov I, Godjevargova T, Mita DG, et al. Electrospun polyacrylonitrile nanofibrous membranes tailored for acetylcholinesterase immobilization. *Journal of Bioactive and Compatable Polymers.* 2010; 25(1): 40–57.

304. Xie J, Hsieh Y. Ultra-high surface fibrous membranes from electrospinning of natural proteins: casein and lipase enzyme. *Journal of Materials Science.* 2003; 38(10): 2125–2133.

305. Herricks TE, Kim S-H, Kim J, Li D, Kwak JH, Grate JW, et al. Direct fabrication of enzyme-carrying polymer nanofibers by electrospinning. *Journal of Materials Chemistry.* 2005; 15(31): 3241–3245.

306. Huang X, Xu Z, Wan L, Innocent C, Seta P. Electro-spun nano fibers modified with phospholipid moieties for enzyme immobilization. *Macromolecular Rapid Communications.* 2006; 27(16): 1341–1345.

307. Wang ZG, Wang JQ, Xu ZK. Immobilization of lipase from Candida rugosa on electrospun polysulfone nanofibrous membranes by adsorption. *Journal of Molecular Catalysis B: Enzymatic.* 2006; 42(1–2): 45–51.

308. Hall EAH. *Chemical Sensors & Biosensors for Medical and Biological Applications*; 1998.

309. Wang X, Lee S, Drew C, Senecal KJ, Kumar J, Samuelson LA. Electrospun nanofibrous membranes for optical sensing. *Polymeric Materials Science and Engineering.* 2001; 85: 617–618.

310. Wang X, Drew C, Lee S-H, Senecal KJ, Kumar J, Samuelson LA. Electrospun nanofibrous membrane for highly sensitive optical sensors. *Nano Letters.* 2002; 2(11): 1273–1275.

311. Wang X, Lee S, Ku B, Samuelson LA, Kumar J. Synthesis and electrospinning of a novel fluorescent polymer PMMA-PM for quenching-based optical sensing. *Journal of Macromolecular Science, Pure and Applied Chemistry.* 2002; A39(10): 1241–1249.

312. Deng C, Gong P, He Q, Cheng J, He C, Shi L, et al. Highly fluorescent TPA-PBPV nanofibers with amplified sensory response to TNT. *Chemical Physics Letters.* 2009; 483(4–6): 219–223.

313. Zhang Y, Li J, An G, He X. Highly porous SnO_2 fibers by electrospinning and oxygen plasma etching and its ethanol-sensing properties. *Sensors and Actuators, B: Chemical.* 2010; 144(1): 43–48.

314. Wang X, Kim Y, Drew C, Ku B, Kumar J, Samuelson LA. Electrostatic assembly of conjugated polymer thin layers on electrospun nanofibrous membranes for biosensors. *Nano Letters.* 2004; 4(2): 331–334.

315. Tao S, Li G, Yin J. Fluorescent nanofibrous membranes for trace detection of TNT vapor. *Journal of Materials Chemistry.* 2007; 17(26): 2730–2736.

316. Yoon J, Chae SK, Kim JM. Colorimetric sensors for volatile organic compounds (VOCs) based on conjugated polymer-embedded electrospun fibers. *Journal of the American Chemical Society.* 2007; 129(11): 3038–3039.

317. Ding B, Yamazaki M, Shiratori S. Electrospun fibrous polyacrylic acid membrane-based gas sensors. *Sensors and Actuators, B: Chemical.* 2005; B106(1): 477–483.

318. Luoh R, Hahn HT. Electrospun nanocomposite fiber mats as gas sensors. *Composites Science and Technology.* 2006; 66(14): 2436–2441.

319. Im JS, Kang S, Lee S, Lee Y. Improved gas sensing of electrospun carbon fibers based on pore structure, conductivity and surface modification. *Carbon.* 2010; 48(9): 2573–2581.

320. Park JY, Asokan K, Choi SW, Kim SS. Growth kinetics of nanograins in SnO2 fibers and size dependent sensing properties. *Sensors and Actuators, B: Chemical.* 2011; B152(2): 254–260.

321. Gouma PI. Nanostructured polymorphic oxides for advanced chemosensors. *Reviews on Advanced Materials Science.* 2003; 5(2): 147–154.

322. Wang G, Ji Y, Huang X, Yang X, Gouma PI, Dudley M. Fabrication and characterization of polycrystalline WO_3 nanofibers and their application for ammonia sensing. *Journal of Physical Chemistry B.* 2006; 110(47): 23777–23782.

323. Sawicka KM, Prasad AK, Gouma PI. Metal oxide nanowires for use in chemical sensing applications. *Sensor Letters.* 2005; 3(1): 31–35.

324. Kim ID, Rothschild A, Lee BH, Kim DY, Jo SM, Tuller HL. Ultrasensitive chemiresistors based on electrospun TiO_2 nanofibers. *Nano Letters.* 2006; 6(9): 2009–2013.

325. Yang M, et al. Fabrication and photoelectric oxygen sensing characteristics of electrospun Co doped ZnO nanofibres. *Applied Physics A, Materials Science and Processing.* 2007; 89: 427–430.

326. Aussawasathien D, Dong JH, Dai L. Electrospun polymer nanofiber sensors. *Synthetic Metals.* 2005; 154(1–3): 37–40.

327. Bishop A, Gouma P. Leuco-emeraldine based polyaniline—poly-vinyl-pyrrolidone electrospun composites and bio-composites: a preliminary study of sensing behavior. *Reviews on Advanced Materials Science.* 2005; 10(3): 209–214.

328. Pinto NJ, Gonzalez R, Johnson AT, Jr., MacDiarmid AG. Electrospun hybrid organic/inorganic semiconductor Schottky nanodiode. *Applied Physics Letters.* 2006; 89(3): 033505/1-/3.

329. Choi J, Lee J, Choi J, Jung D, Shim SE. Electrospun PEDOT:PSS/PVP nanofibers as the chemiresistor in chemical vapour sensing. *Synthetic Metals.* 2010; 160(13–14): 1415–1421.

330. Laxminarayana K, Jalili N. Functional nanotube-based textiles: pathway to next generation fabrics with enhanced sensing capabilities. *Textile Research Journal.* 2005; 75(9): 670–680.

331. Kessick R, Tepper G. Electrospun polymer composite fiber arrays for the detection and identification of volatile organic compounds. *Sensors and Actuators, B: Chemical.* 2006; B117(1): 205–210.

332. Qi Q, Zhang T, Wang S, Zheng X. Humidity sensing properties of KCl-doped ZnO nanofibers with super-rapid response and recovery. *Sensors and Actuators, B: Chemical.* 2009; B137(2): 649–655.

333. Qi Q, Feng Y, Zhang T, Zheng X, Lu G. Influence of crystallographic structure on the humidity sensing properties of KCl-doped TiO_2 nanofibers. *Sensors and Actuators, B: Chemical.* 2009; B139(2): 611–617.

334. He Y, Zhang T, Zheng W, Wang R, Liu X, Xia Y, et al. Humidity sensing properties of $BaTiO_3$ nanofiber prepared via electrospinning. *Sensors and Actuators, B: Chemical.* 2010; 146(1): 98–102.

335. Manesh KM, Santhosh P, Gopalan A, Lee KP. Electrospun poly(vinylidene fluoride)/poly(aminophenylboronic acid) composite nanofibrous membrane as a novel glucose sensor. *Analytical Biochemistry.* 2007; 360(2): 189–195.

336. Song MY, Kim DK, Ihn KJ, Jo SM, Kim DY. Electrospun TiO_2 electrodes for dye-sensitized solar cells. *Nanotechnology.* 2004; 15(12): 1861–1865.

337. Jo SM, Song MY, Ahn YR, Park CR, Kim DY. Nanofibril formation of electrospun TiO_2 fibers and its application to dye-sensitized slar cells. *Journal of Macromolecular Science, Part A: Pure and Applied Chemistry.* 2005; 42(11): 1529–1540.

338. Priya ARS, Subramania A, Jung Y-S, Kim K-J. High-performance quasi-solid-state dye-sensitized solar cell based on an electrospun PVdF-HFP membrane electrolyte. *Langmuir.* 2008; 24(17): 9816–9819.

339. Park SH, Kim JU, Lee SY, Lee WK, Lee JK, Kim MR. Dye-sensitized solar cells using polymer electrolytes based on poly(vinylidene fluoride-hexafluoro propylene) nanofibers by electrospinning method. *Journal of Nanoscience and Nanotechnology.* 2008; 8(9): 4889–4894.

340. Onozuka K, Ding B, Tsuge Y, Naka T, Yamazaki M, Sugi S, et al. Electrospinning processed nanofibrous TiO_2 membranes for photovoltaic applications. *Nanotechnology.* 2006; 17(4): 1026–1031.

341. Kim ID, Hong JM, Lee BH, Kim DY, Jeon EK, Choi DK, et al. Dye-sensitized solar cells using network structure of electrospun ZnO nanofiber mats. *Applied Physics Letters.* 2007; 91(16): 163109/1-/3.

342. Zhang W, Zhu R, Liu X, Liu B, Ramakrishna S. Facile construction of nanofibrous ZnO photoelectrode for dye-sensitized solar cell applications. *Applied Physics Letters.* 2009; 95(4): 043304/1-/3.

343. Song MY, Kim DK, Jo SM, Kim DY. Enhancement of the photocurrent generation in dye-sensitized solar cell based on electrospun TiO_2 electrode by surface treatment. *Synthetic Metals.* 2005; 155(3): 635–638.

344. Lee BH, Song MY, Jang SY, Jo SM, Kwak SY, Kim DY. Charge transport characteristics of high efficiency dye-sensitized solar sells based on electrospun TiO_2 nanorod photoelectrodes. *Journal of Physical Chemistry C.* 2009; 113(51): 21453–21457.

345. Sudhagar P, Jung JH, Park S, Lee Y-G, Sathyamoorthy R, Kang YS, et al. The performance of coupled (CdS:CdSe) quantum dot-sensitized TiO_2 nanofibrous solar cells. *Electrochemistry Communications.* 2009; 11(11): 2220–2224.

346. Yang S, Nair AS, Jose R, Ramakrishna S. Electrospun TiO_2 nanorods assembly sensitized by CdS quantum dots: a low-cost photovoltaic material. *Energy and Environmental Science.* 2010; 3(12): 2010–2014.

347. Archana PS, Jose R, Jin T, Vijila C, Yusoff MM, Ramakrishna S. Structural and electrical properties of Nb-doped anatase TiO_2 nanowires by electrospinning. *Journal of the American Ceramic Society.* 2010; 93(12): 4096–4102.

348. Wu H, Hu L, Rowell MW, Kong D, Cha JJ, McDonough JR, et al. Electrospun metal nanofiber webs as high-performance transparent electrode. *Nano Letters.* 2010; 10(10): 4242–4248.

349. Joshi P, Zhang L, Chen Q, Galipeau D, Fong H, Qiao Q. Electrospun carbon nanofibers as low-cost counter electrode for dye-sensitized solar cells. *ACS Applied Materials and Interfaces.* 2010; 2(12): 3572–3577.

350. Sundmacher K. Fuel cell engineering: Toward the design of efficient electrochemical power plants. *Industrial and Engineering Chemistry Research.* 2010; 49(21): 10159–10182.

351. Kim YS, Nam SH, Shim HS, Ahn HJ, Anand M, Kim WB. Electrospun bimetallic nanowires of PtRh and PtRu with compositional variation for methanol electrooxidation. *Electrochemistry Communications.* 2008; 10(7): 1016–1019.

352. Kim HJ, Kim YS, Seo MH, Choi SM, Kim WB. Pt and PtRh nanowire electrocatalysts for cyclohexane-fueled polymer electrolyte membrane fuel cell. *Electrochemistry Communications.* 2009; 11(2): 446–449.

353. Li M, Han G, Yang B. Fabrication of the catalytic electrodes for methanol oxidation on electrospinning-derived carbon fibrous mats. *Electrochemistry Communications.* 2008; 10(6): 880–883.

354. Xuyen NT, Jeong HK, Kim G, So KP, An KH, Lee YH. Hydrolysis-induced immobilization of $Pt(acac)_2$ on polyimide-based carbon nanofiber mat and formation of Pt nanoparticles. *Journal of Materials Chemistry.* 2009; 19(9): 1283–1288.

355. Nguyen TX, Kim TH, Geng H-Z, Lee IH, Kim KK, Lee YH. Three-dimensional architecture of carbon nanotube-anchored polymer nanofiber composite. *Journal of Materials Chemistry.* 2009; 19(42): 7822–7825.

356. Lee KM, Choi J, Wycisk R, Pintauro PN, Mather PT. Nafion nanofiber membranes. *ECS Transctions.* 2009; 25(1): 1451–1458.

357. Lin HL, Wang SH, Chiu CK, Yu TL, Chen LC, Huang CC, et al. Preparation of Nafion/poly(vinyl alcohol) electro-spun fiber composite membranes for direct methanol fuel cells. *Journal of Membrane Science.* 2010; 365(1–2): 114–122.

358. Molla S, Compan V. Performance of composite Nafion/PVA membranes for direct methanol fuel cells. *Journal of Power Sources.* 2011; 196(5): 2699–2708.

359. Molla S, Compan V. Polyvinyl alcohol nanofiber reinforced Nafion membranes for fuel cell applications. *Journal of Membrane Science.* 2011; 372(1–2): 191–200.

360. Wang ZL, Song J. Piezoelectric nanogenerators based on zinc oxide nanowire arrays. *Science.* 2006; 312(5771): 243–246.

361]Wang ZL. Self-powered nanotech. *Scientific American.* 2008.

362. Lu MY, Song JH, Lu MP, Lee CY, Chen LJ, Wang ZL. ZnO-ZnS heterojunction and ZnS nanowire arrays for electricity generation. *ACS Nano.* 2009; 3(20): 357–362.

363. Huang CT, Song JH, Tsai CM, Lee WF, Lien DH, Gao ZY, et al. Single-InN-nanowire nanogenerator with upto 1 V output voltage. *Advanced Materials.* 2010; 22(36): 4008–4013.

364. Chang C, Tran VH, Wang J, Fuh Y-K, Lin L. Direct-write piezoelectric polymeric nanogenerator with high energy conversion efficiency. *Nano Letters.* 2010; 10(2): 726–731.

365. Chen X, Xu S, Yao N, Shi Y. 1.6 V nanogenerator for mechanical energy harvesting using PZT nanofibers. *Nano Letters.* 2010; 10(6): 2133–2137.

366. Hansen BJ, Liu Y, Yang R, Wang ZL. Hybrid nanogenerator for concurrently harvesting biomechanical and biochemical energy. *ACS Nano.* 2010; 4(7): 3647–3652.

367. Fang J, Wang X, Lin T. Electrical power generator from randomly oriented electrospun poly(vinylidene fluoride) nanofibre membranes. *Journal of Materials Chemistry.* 2011; 21(30): 11088–11091.

368. Mandal D, Yoon S, Kim KJ. Origin of Piezoelectricity in an Electrospun Poly(vinylidene fluoride-trifluoroethylene) Nanofiber Web-Based Nanogenerator and Nano-Pressure Sensor. *Macromolecular Rapid Communications.* 2011; 32(11): 831–837.

369. Arora P, Zhang Z. Battery separators. *Chemical Reviews* 2004; 104(10): 4419–4462.

370. Kim JR, Choi SW, Jo SM, Lee WS, Kim BC. Electrospun PVdF-based fibrous polymer electrolytes for lithium ion polymer batteries. *Electrochimica Acta.* 2004; 50(1): 69–75.

371. Choi S-S, Lee YS, Joo CW, Lee SG, Park JK, Han K-S. Electrospun PVDF nanofiber web as polymer electrolyte or separator. *Electrochimica Acta.* 2004; 50(2–3): 339–343.

372. Gao K, Hu X, Dai C, Yi T. Crystal structures of electrospun PVDF membranes and its separator application for rechargeable lithium metal cells. *Materials Science and Engineering, B: Solid-State Materials for Advanced Technology.* 2006; 131(1–3): 100–105.

373. Choi SW, Kim JR, Jo SM, Lee WS, Kim YR. Electrochemical and spectroscopic properties of electrospun PAN-based fibrous polymer electrolytes. *Journal of the Electrochemical Society.* 2005; 152(5): A989–A995.

374. Kim JK, Cheruvally G, Li X, Ahn JH, Kim KW, Ahn HJ. Preparation and electrochemical characterization of electrospun, microporous membrane-based composite polymer electrolytes for lithium batteries. *Journal Power Sources.* 2008; 178(2): 815–820.

375. Gopalan AI, Lee K-P, Manesh KM, Santhosh P. Poly(vinylidene fluoride)-polydiphenylamine composite electrospun membrane as high-performance polymer electrolyte for lithium batteries. *Journal of Membrane Science.* 2008; 318(1–2): 422–428.

376. Ding Y, Zhang P, Long Z, Jiang Y, Xu F, Di W. The ionic conductivity and mechanical property of electrospun P(VdF-HFP)/PMMA membranes for lithium ion batteries. *Journal of Membrane Science.* 2009; 329(1–2): 56–59.

377. Raghavan P, Zhao X, Shin C, Baek D, Choi J, Manuel J, et al. Preparation and electrochemical characterization of polymer electrolytes based on electrospun poly(vinylidene fluoride-co-hexafluoropropylene)/polyacrylonitrile blend/composite membranes for lithium batteries. *Journal of Power Sources.* 2010; 195(18): 6088–6094.

378. Ahn YC, Park SK, G.T. K, Hwang YJ, Lee CG, Shin HS, et al. Development of high efficiency nanofilters made of nanofibers. *Current Applied Physics.* 2006; 6: 1030–1035.

379. Wang L, Yu Y, Chen PC, Zhang DW, Chen CH. Electrospinning synthesis of C/Fe3O4 composite nanofibers and their application for high performance lithium-ion batteries. *Journal of Power Sources.* 2008; 183(2): 717–723.

380. Yu Y, Gu L, Wang C, Dhanabalan A, van APA, Maier J. Encapsulation of Sn@carbon nanoparticles in bamboo-like hollow carbon nanofibers as an anode material in lithium-based batteries. *Angewandte Chemie, International Edition.* 2009; 48(35): 6485-9, S/1-S/3.

381. Ding Y, Zhang P, Long Z, Jiang Y, Gao D. The morphology, structure and electrochemical properties of LiNi1/3Mn1/3Co1/3O2 prepared by electrospun method. *Journal of Alloys and Compounds.* 2008; 462(1–2): 340–342.

382. Mai L, Xu L, Han C, Xu X, Luo Y, Zhao S, et al. Electrospun ultralong hierarchical vanadium oxide nanowires with high performance for lithium ion batteries. *Nano Letters.* 2010; 10(11): 4750–4755.

383. Lee SH, Jung MJ, Im JS, Sheem KY, Lee YS. Preparation and characterization of electrospun $LiFePO_4$/carbon complex improving rate performance at high C-rate. *Research on Chemical Intermediates.* 2010; 36(6–7): 591–602.

384. Kim C, Yang KS. Electrochemical properties of carbon nanofiber web as an electrode for supercapacitor prepared by electrospinning. *Applied Physics Letters.* 2003; 83(6): 1216–1218.

385. Kim C, Park SH, Lee WJ, Yang KS. Characteristics of supercapacitor electrodes of PBI-based carbon nanofiber web prepared by electrospinning. *Electrochimica Acta.* 2004; 50(2–3): 877–881.

386. Kim C, Ngoc BTN, Yang KS, Kojima M, Kim YA, Kim YJ, et al. Self-sustained thin webs consisting of porous carbon nanofibers for supercapacitors via the electrospinning of polyacrylonitrile solutions containing zinc chloride. *Advanced Materials.* 2007; 19(17): 2341–2346.

387. Guo Q, Zhou X, Li X, Chen S, Seema A, Greiner A, et al. Supercapacitors based on hybrid carbon nanofibers containing multiwalled carbon nanotubes. *Journal of Materials Chemistry.* 2009; 19(18): 2810–2816.

388. Li J, Liu E-h, Li W, Meng X-y, Tan S-t. Nickel/carbon nanofibers composite electrodes as supercapacitors prepared by electrospinning. *Journal of Alloys and Compounds.* 2009; 478(1–2): 371–374.

389. Niu H, Zhang J, Xie Z, Wang X, Lin T. Preparation, structure and supercapacitance of bonded carbon nanofiber electrode materials. *Carbon.* 2011; 49(7): 2380–2388.

390. Choi S-H, Hyun T-S, Lee H, Jang S-Y, Oh S-G, Kim I-D. Facile synthesis of highly conductive platinum nanofiber mats as conducting core for high rate redox supercapacitor. *Electrochemical and Solid-State Letters.* 2010; 13(6): A65–A68.

391. Wee G, Soh HZ, Cheah YL, Mhaisalkar SG, Srinivasan M. Synthesis and electrochemical properties of electrospun V2O5 nanofibers as supercapacitor electrodes. *Journal of Materials Chemistry.* 2010; 20(32): 6720–6725.

392. Jones R, Thomas G. An overview of materials for the hydrogen economy. *JOM Journal of the Minerals, Metals and Materials Society.* 2007; 59(12): 50–55.

393. Mandal TK, Gregory DH. Hydrogen storage materials: present scenarios and future directions. *Annual Reports Section "A"* (Inorganic Chemistry). 2009; 105: 21–54.

394. Kim DK, Park SH, Kim BC, Chin BD, Jo SM, Kim DY. Electrospun polyacrylonitrile-based carbon nanofibers and their hydrogen storage. *Macromolecular Research.* 2005; 13(6): 521–528.

395. Hong S, Kim D, Jo S, Kim D, Chin B, Lee D. Graphite nanofibers prepared from catalytic graphitization of electrospun poly(vinylidene fluoride) nanofibers and their hydrogen storage capacity. *Catalysis Today.* 2007; 120(3–4): 413–419.

396. Srinivasan SS, Ratnadurai R, Niemann MU, Phani AR, Goswami DY, Stefanakos EK. Reversible hydrogen storage in electrospun polyaniline fibers. *International Journal of Hydrogen Energy.* 2010; 35(1): 225–230.

397. Rose M, Kockrick E, Senkovska I, Kaskel S. High surface area carbide-derived carbon fibers produced by electrospinning of polycarbosilane precursors. *Carbon.* 2010; 48(2): 403–407.

398. Im JS, Park SJ, Lee YS. Superior prospect of chemically activated electrospun carbon fibers for hydrogen storage. *Materials Research Bulletin.* 2009; 44(9): 1871–1878.

399. Im JS, Park SJ, Kim T, Lee YS. Hydrogen storage evaluation based on investigations of the catalytic properties of metal/metal oxides in

electrospun carbon fibers. *International Journal of Hydrogen Energy.* 2009; 34(8): 3382–3388.

400. Im J, Park S, Lee Y. The metal-carbon-fluorine system for improving hydrogen storage by using metal and fluorine with different levels of electronegativity. *International Journal of Hydrogen Energy.* 2009; 34(3): 1423–1428.

401. Im JS, Kwon O, Kim Y, Park S, Lee Y. The effect of embedded vanadium catalyst on activated electrospun CFs for hydrogen storage. *Microporous and Mesoporous Materials.* 2008; 115(3): 514–521.

402. Im JS, Park SJ, Kim TJ, Kim YH, Lee YS. The study of controlling pore size on electrospun carbon nanofibers for hydrogen adsorption. *Journal of Colloid and Interface Science.* 2008; 318(1): 42–49.

403. Bergshoef MM, Vancso GJ. Transparent nanocomposites with ultrathin, electrospun Nylon-4,6 fiber reinforcement. *Advanced Materials.* 1999; 11(16): 1362–1365.

404. Li G, Li P, Zhang C, Yu Y, Liu H, Zhang S, et al. Inhomogeneous toughening of carbon fiber/epoxy composite using electrospun polysulfone nanofibrous membranes by in situ phase separation. *Composites Science and Technology.* 2008; 68(3–4): 987–994.

405. Zhang J, Lin T, Wang X. Electrospun nanofibre toughened carbon/epoxy composites: effects of polyetherketone cardo (PEK-C) nanofibre diameter and interlayer thickness. *Composites Science and Technology.* 2010; 70(11): 1660–1666.

406. Zhang J, Wang X, Lin T. Synergistic effects of polyetherketone cardo (PEK-C)/carbon nanofibre composite on eposy resins. Composites Science and Technology. 2011: accepted for publication (18 Mar 2011), DIO:10.1016/j.compscitech.2011.03.008.

407. Kim JS, Eneker DH. Mechanical properties of composites using ultrafine electrospun fibers. *Polymer Composites.* 1999; 20(1): 124–131.

408. Fong H. Electrospun nylon 6 nanofiber reinforced BIS-GMA/TEGDMA dental restorative composite resins. *Polymer.* 2004; 45(7): 2427–2432.

409. Pinto NJ, Johnson AT, Jr., MacDiarmid AG, Mueller CH, Theofylaktos N, Robinson DC, et al. Electrospun polyaniline/polyethylene oxide nanofiber field-effect transistor. *Applied Physics Letters.* 2003; 83(20): 4244–4246.

410. Yang S, Wang C, Chen S. A release-induced response for the rapid recognition of latent fingerprints and formation of inkjet-printed patterns. *Angewandte Chemie, International Edition.* 2011; 50(16): 3706–3709.

411. Varesano A, Carletto RA, Mazzuchetti G. Experimental investigations on the multi-jet electrospinning process. *Journal of Materials Processing Technology.* 2009; 209(11): 5178–5185.

412. Yang E, Shi J, Xue Y. Influence of electric field interference on double nozzles electrospinning. *Journal of Applied Polymer Science.* 2010; 116(6): 3688–3692.

413. Kim G, Cho Y-S, Kim WD. Stability analysis for multi-jets electrospinning process modified with a cylindrical electrode. *European Polymer Journal.* 2006; 42(9): 2031–2038.

414. Theron SA, Yarin AL, Zussman E, Kroll E. Multiple jets in electrospinning: experiment and modeling. *Polymer.* 2005; 46(9): 2889–2899.

415. Dosunmu OO, Chase GG, Kataphinan W, Reneker DH. Electrospinning of polymer nanofibres from multiple jets on a porous tubular surface. *Nanotechnology.* 2006; 17(4): 1123–1127.

416. Varabhas JS, Chase GG, Reneker DH. Electrospun nanofibers from a porous hollow tube. *Polymer.* 2008; 49(19): 4226–4229.

417. Wang X, Niu H, Lin T, Wang X. Needleless electrospinning of nanofibers with a conical wire coil. *Polymer Engineering and Science.* 2009; 49(8): 1582–1586.

418. Thoppey NM, Bochinski JR, Clarke LI, Gorga RE. Unconfined fluid electrospun into high quality nanofibers from a plate edge. *Polymer.* 2010; 51(21): 4928–4936.

419. Tang S, Zeng Y, Wang X. Splashing needleless electrospinning of nanofibers. *Polymer Engineering and Science.* 2010; 50(11): 2252–2257.

420. Lu B, Wang Y, Liu Y, Duan H, Zhou J, Zhang Z, et al. Superhigh-Throughput Needleless Electrospinning Using a Rotary Cone as Spinneret. *Small.* 2010; 6(15): 1612–1616.

421. Yarin AL, Zussman E. Upward needleless electrospinning of multiple nanofibers. *Polymer.* 2004; 45(9): 2977–2980.

422. Liu Y, He J-H, Yu J-Y. Bubble-electrospinning: a novel method for making nanofibers. *Journal of Physics: Conference Series.* 2008; 96(1): 012001.

423. Jirsak O, Sanetrnik F, Lukas D, Kotek V, Martinova L, Chaloupek J. A method of nanofibres production from a polymer solution using electrostatic spinning and a device for carrying out the method. WO 2005/024101 A1. 2005.

424. Lukas D, Sarkar A, Pokorny P. Self-organization of jets in electrospinning from free liquid surface: A generalized approach. *Journal of Applied Physics.* 2008; 103: 084309.

425. Niu H, Lin T, Wang X. Needleless electrospinning. I. A comparison of cylinder and disk nozzles. *Journal of Applied Polymer Science.* 2009; 114(6): 3524–3530.

426. Lin T, Wang X, Wang X, Niu H. Electrostatic spinning assembly. WO/2010/043002. 2010.

427. Wang X, Niu H, Wang X, Lin T. Large-scale electrospinning of polymer nanofibers using needleless nozzle. *Proceedings of the 38th Textile Research Symposium.* 2009; 117–122.

428. Wu D, Huang X, Lai X, Sun D, Lin L. High throughput tip-less electrospinning via a circular cylindrical electrode. *Journal of Nanoscience and Nanotechnology.* 2010; 10(7): 4221–4226.

429. Cengiz F, Jirsak O. The effect of salt on the roller electrospinning of polyurethane nanofibers. *Fibers and Polymers.* 2009; 10(2): 177–184.

430. Lin T, Fang J, Wang H, Wang X. Using chitosan as a thickener for electrospinning dilute PVA solutions to improve fibre uniformity. *Nanotechnology.* 2006; 17: 3718–3723.

431. Li ZH, Wang H. Ce Electrostatic forces induce poly(vinyl alcohol)-protected copper nanoparticles to form copper/poly(vinyl alcohol) nanocables via electrospinning. *Macromolecular Rapid Communications.* 2006; 27(2): 152–155.

432. Hong Y, Shang T, Li Y, Wang L, Wang C, Chen X, et al. Synthesis using electrospinning and stabilization of single layer macroporous films and fibrous networks of poly(vinyl alcohol). *Journal of Membrane Science.* 2006; 276(1–2): 1–7.

433. Lin T, Fang J, Wang H, Wang X. Using chitosan as a thickener for electrospinning dilute PVA solutions to improve fibre uniformity. *Nanotechnology.* 2006; 17: 3718.

434. Taylor G. Electrically driven jets. *Proceedings of the Royal Society of London Series A, Mathematical and Physical Sciences.* 1969; 313(1515): 453–475.

435. Jirsak O, Sysel P, Sanetrnik F, Hruza J, Chaloupek J. Polyamic acid nanofibers produced by needleless electrospinning. *Journal of Nanomaterials.* 2010; 2010: 1–7.

Index